BRICKLAYING

LEVEL 2 DIPLOMA

JOHN CARRUTHERS

IAN COOTE

Leeds
College of
Building

Nelson Thornes

Published in 2013 by:
Nelson Thornes Ltd
Delta Place
27 Bath Road
CHELTENHAM
GL53 7TH
United Kingdom

13 14 15 16 17 / 10 9 8 7 6 5 4 3 2 1

A catalogue record for this book is available from the British Library

ISBN 978 1 4085 2123 6

Cover photograph: DNY59/iStockphoto

Page make-up by GreenGate Publishing Services, Tonbridge, Kent

Printed in Croatia by Zrinski

Note to learners and tutors

This book clearly states that a risk assessment should be undertaken and the correct PPE worn for the particular activities before any practical activity is carried out. Risk assessments were carried out before photographs for this book were taken and the models are wearing the PPE deemed appropriate for the activity and situation. This was correct at the time of going to print. Colleges may prefer that their learners wear additional items of PPE not featured in the photographs in this book and should instruct learners to do so in the standard risk assessments they hold for activities undertaken by their learners. Learners should follow the standard risk assessments provided by their college for each activity they undertake which will determine the PPE they wear.

CONTENTS

Introduction   iv

Contributors to this book   v

1 Health, Safety and Welfare in Construction and Associated Industries   **1**

2 Understand Information, Quantities and Communication with Others   **39**

3 Understanding Construction Technology   **67**

4 Set Out Masonry Structures   **95**

5 Construct Solid Walling Incorporating Piers and Arches   **127**

6 Construct Cavity Walling Forming Masonry Structures   **177**

7 Construct Masonry Cladding   **213**

8 Construct Thin Joint Masonry   **223**

Index   245

Acknowledgements   250

INTRODUCTION

About this book

This book has been written for the Cskills Awards Level 2 Diploma in Bricklaying. It covers all the units of the qualification, so you can feel confident that your book fully covers the requirements of your course.

This book contains a number of features to help you acquire the knowledge you need. It also demonstrates the practical skills you will need to master to successfully complete your qualification. We've included additional features to show how the skills and knowledge can be applied to the workplace, as well as tips and advice on how you can improve your chances of gaining employment.

The features include:

* chapter openers which list the learning outcomes you must achieve in each unit

* key terms that provide explanations of important terminology that you will need to know and understand

* Did you know? margin notes to provide key facts that are helpful to your learning

* practical tips to explain facts or skills to remember when undertaking practical tasks

* Reed tips to offer advice about work, building your CV and how to apply the skills and knowledge you have learnt in the workplace

* case studies that are based on real tradespeople who have undertaken apprenticeships and explain why the skills and knowledge you learn with your training provider are useful in the workplace

* practical tasks that provide step-by-step directions and illustrations for a range of projects you may do during your course

* Test yourself multiple choice questions that appear at the end of each unit to give you the chance to revise what you have learnt and to practise your assessment (your tutor will give you the answers to these questions).

Further support for this book can be found at our website, www.planetvocational.com/subjects/build

KEY TERMS

DID YOU KNOW?

PRACTICAL TIP

REED TIP

CASE STUDY

PRACTICAL TASK

TEST YOURSELF

Planet Vocational

CONTRIBUTORS TO THIS BOOK

Reed Property & Construction

Reed Property & Construction specialises in placing staff at all levels, in both temporary and permanent positions, across the complete lifecycle of the construction process. Our consultants work with most major construction companies in the UK and our clients are involved with the design, build and maintenance of infrastructure projects throughout the UK.

Expert help

As a leading recruitment consultancy for mid–senior level construction staff in the UK, Reed Property & Construction is ideally placed to advise new workers entering the sector, from building a CV to providing expertise and sharing our extensive sector knowledge with you. That's why, throughout this book, you will find helpful hints from our highly experienced consultants, all designed to help you find that first step on the construction career ladder. These tips range from advice on CV writing to interview tips and techniques, and are all linked in with the learning material in this book.

Work-related advice

Reed Property & Construction has gained insights from some of our biggest clients – leading recruiters within the industry – to help you understand the mind-set of potential employers. This includes the traits and skills that they would like to see in their new employees, why you need the skills taught in this book and how they are used on a day to day basis within their organisations.

Getting your first job

This invaluable information is not available anywhere else and is all geared towards helping you gain a position with an employer once you've completed your studies. Entry level positions are not usually offered by recruitment companies, but the advice we've provided will help you to apply for jobs in construction and hopefully gain your first position as a skilled worker.

CONTRIBUTORS TO THIS BOOK

The case studies in this book feature staff from Laing O'Rourke and South Tyneside Homes.

Laing O'Rourke is an international engineering company that constructs large-scale building projects all over the world. Originally formed from two companies, John Laing (founded in 1848) and R O'Rourke and Son (founded in 1978) joined forces in 2001.

At Laing O'Rourke, there is a strong and unique apprenticeship programme. It runs a four-year 'Apprenticeship Plus' scheme in the UK, combining formal college education with on-the-job training. Apprentices receive support and advice from mentors and experienced tradespeople, and are given the option of three different career pathways upon completion: remaining on site, continuing into a further education programme, or progressing into supervision and management.

The company prides itself on its people development, supporting educational initiatives and investing in its employees. Laing O'Rourke believes in collaboration and teamwork as a path to achieving greater success, and strives to maintain exceptionally high standards in workplace health and safety.

South Tyneside Homes was launched in 2006, and was previously part of South Tyneside Council. It now works in partnership with the council to repair and maintain 18,000 properties within the borough, including delivering parts of the Decent Homes Programme.

South Tyneside Homes believes in putting back into the community, with 90 per cent of its employees living in the borough itself. Equality and diversity, as well as health and wellbeing of staff, is a top priority, and it has achieved the Gold Status Investors in People Award.

South Tyneside Homes is committed to the development of its employees, providing opportunities for further education and training and great career paths within the company – 80 per cent of its management team started as apprentices with the company. As well as looking after its staff and their community, the company looks after the environment too, running a renewable energy scheme for council tenants in order to reduce carbon emissions and save tenants money.

The apprenticeship programme at South Tyneside Homes has been recognised nationally, having trained over 80 young people in five main trade areas over the past six years. One of the UK's Top 100 Apprenticeship Employers, it is an Ambassador on the panel of the National Apprentice Service. It has won the Large Employer of the Year Award at the National Apprenticeship Awards and several of its apprentices have been nominated for awards, including winning the Female Apprentice of the Year for the local authority.

South Tyneside Council's Housing Company

Chapter 1

Unit CSA–L1Core01

HEALTH, SAFETY AND WELFARE IN CONSTRUCTION AND ASSOCIATED INDUSTRIES

LEARNING OUTCOMES

LO1: Know the health and safety regulations, roles and responsibilities

LO2: Know the accident and emergency procedures and how to report them

LO3: Know how to identify hazards on construction sites

LO4: Know about health and hygiene in a construction environment

LO5: Know how to handle and store materials and equipment safely

LO6: Know about basic working platforms and access equipment

LO7: Know how to work safely around electricity in a construction environment

LO8: Know how to use personal protective equipment (PPE) correctly

LO9: Know the fire and emergency procedures

LO10: Know about signs and safety notices

INTRODUCTION

The aim of this chapter is to:

* help you to source relevant safety information

* help you to use the relevant safety procedures at work.

HEALTH AND SAFETY REGULATIONS, ROLES AND RESPONSIBILITIES

The construction industry can be dangerous, so keeping safe and healthy at work is very important. If you are not careful, you could injure yourself in an accident or perhaps use equipment or materials that could damage your health. Keeping safe and healthy will help ensure that you have a long and injury-free career.

Although the construction industry is much safer today than in the past, more than 2,000 people are injured and around 50 are killed on site every year. Many others suffer from long-term ill-health such as deafness, spinal damage, skin conditions or breathing problems.

Key health and safety legislation

Laws have been created in the UK to try to ensure safety at work. Ignoring the rules can mean injury or damage to health. It can also mean losing your job or being taken to court.

The two main laws are the Health and Safety at Work etc. Act **(HASAWA)** and the Control of Substances Hazardous to Health Regulations **(COSHH)**.

The Health and Safety at Work etc. Act (HASAWA) (1974)

This law applies to all working environments and to all types of worker, sub-contractor, employer and all visitors to the workplace. It places a duty on everyone to follow rules in order to ensure health, safety and welfare. Businesses must manage health and safety risks, for example by providing appropriate training and facilities. The Act also covers first aid, accidents and ill health.

Reporting of Injuries, Diseases and Dangerous Occurrences Regulations (RIDDOR) (1995)

Under RIDDOR, employers are required to report any injuries, diseases or dangerous occurrences to the **Health and Safety Executive (HSE)**. The regulations also state the need to maintain an **accident book**.

Control of Substances Hazardous to Health (COSHH) (2002)

In construction, it is common to be exposed to substances that could cause ill health. For example, you may use oil-based paints or preservatives, or work in conditions where there is dust or bacteria.

Employers need to protect their employees from the risks associated with using hazardous substances. This means assessing the risks and deciding on the necessary precautions to take.

Any control measures (things that are being done to reduce the risk of people being hurt or becoming ill) have to be introduced into the workplace and maintained; this includes monitoring an employee's exposure to harmful substances. The employer will need to carry out health checks and ensure that employees are made aware of the dangers and are supervised.

Control of Asbestos at Work Regulations (2012)

Asbestos was a popular building material in the past because it was a good insulator, had good fire protection properties and also protected metals against corrosion. Any building that was constructed before 2000 is likely to have some asbestos. It can be found in pipe insulation, boilers and ceiling tiles. There is also asbestos cement roof sheeting and there is a small amount of asbestos in decorative coatings such as Artex.

Asbestos has been linked with lung cancer, other damage to the lungs and breathing problems. The regulations require you and your employer to take care when dealing with asbestos:

* You should always assume that materials contain asbestos unless it is obvious that they do not.

* A record of the location and condition of asbestos should be kept.

* A risk assessment should be carried out if there is a chance that anyone will be exposed to asbestos.

The general advice is as follows:

* Do not remove the asbestos. It is not a hazard unless it is removed or damaged.

* Remember that not all asbestos presents the same risk. Asbestos cement is less dangerous than pipe insulation.

* Call in a specialist if you are uncertain.

Provision and Use of Work Equipment Regulations (PUWER) (1998)

PUWER concerns health and safety risks related to equipment used at work. It states that any risks arising from the use of equipment must either be prevented or controlled, and all suitable safety measures must have been taken. In addition, tools need to be:

* suitable for their intended use

* safe

REED TIP

Employers will want to know that you understand the importance of health and safety. Make sure you know the reasons for each safe working practice.

* well maintained

* used only by those who have been trained to do so.

Manual Handling Operations Regulations (1992)

These regulations try to control the risk of injury when lifting or handling bulky or heavy equipment and materials. The regulations state as follows:

* Hazardous manual handling should be avoided if possible.

* An assessment of hazardous manual handling should be made to try to find alternatives.

* You should use mechanical assistance where possible.

* The main idea is to look at how manual handling is carried out and finding safer ways of doing it.

Personal Protection at Work Regulations (PPE) (1992)

This law states that employers must provide employees with personal protective equipment **(PPE)** at work whenever there is a risk to health and safety. PPE needs to be:

* suitable for the work being done

* well maintained and replaced if damaged

* properly stored

* correctly used (which means employees need to be trained in how to use the PPE properly).

Work at Height Regulations (2005)

Whenever a person works at any height there is a risk that they could fall and injure themselves. The regulations place a duty on employers or anyone who controls the work of others. This means that they need to:

* plan and organise the work

* make sure those working at height are **competent**

* assess the risks and provide appropriate equipment

* manage work near or on fragile surfaces

* ensure equipment is inspected and maintained.

In all cases the regulations suggest that, if it is possible, work at height should be avoided. Perhaps the job could be done from ground level? If it is not possible, then equipment and other measures are needed to prevent the risk of falling. When working at height measures also need to be put in place to minimise the distance someone might fall.

Figure 1.1 Examples of personal protective equipment

Employer responsibilities under HASAWA

HASAWA states that employers with five or more staff need their own health and safety policy. Employers must assess any risks that may be involved in their workplace and then introduce controls to reduce these risks. These risk assessments need to be reviewed regularly.

Employers also need to supply personal protective equipment (PPE) to all employees when it is needed and to ensure that it is worn when required.

Specific employer responsibilities are outlined in Table 1.1.

Employee responsibilities under HASAWA

HASAWA states that all those operating in the workplace must aim to work in a safe way. For example, they must wear any PPE provided and look after their equipment. Employees should not be charged for PPE or any actions that the employer needs to take to ensure safety.

Specific employer responsibilities are outlined in Table 1.1. Table 1.2 identifies the key employee responsibilities.

KEY TERMS

Risk

– the likelihood that a person may be harmed if they are exposed to a hazard.

Hazard

– a potential source of harm, injury or ill-health.

Near miss

– any incident, accident or emergency that did not result in an injury but could have done so.

Employer responsibility	Explanation
Safe working environment	Where possible all potential risks and hazards should be eliminated.
Adequate staff training	When new employees begin a job their induction should cover health and safety. There should be ongoing training for existing employees on risks and control measures.
Health and safety information	Relevant information related to health and safety should be available for employees to read and have their own copies.
Risk assessment	Each task or job should be investigated and potential risks identified so that measures can be put in place. A risk assessment and method statement should be produced. The method statement will tell you how to carry out the task, what PPE to wear, equipment to use and the sequence of its use.
Supervision	A competent and experienced individual should always be available to help ensure that health and safety problems are avoided.

Table 1.1 Employer responsibilities under HASAWA

Employee responsibility	Explanation
Working safely	Employees should take care of themselves, only do work that they are competent to carry out and remove obvious hazards if they are seen.
Working in partnership with the employer	Co-operation is important and you should never interfere with or misuse any health and safety signs or equipment. You should always follow the site rules.
Reporting hazards, near misses and accidents correctly	Any health and safety problems should be reported and discussed, particularly a near miss or an actual accident.

Table 1.2 Employee responsibilities under HASAWA

Health and Safety Executive

The Health and Safety Executive (HSE) is responsible for health, safety and welfare. It carries out spot checks on different workplaces to make sure that the law is being followed.

HSE inspectors have access to all areas of a construction site and can also bring in the police. If they find a problem then they can issue an **improvement notice**. This gives the employer a limited amount of time to put things right.

In serious cases, the HSE can issue a **prohibition notice**. This means all work has to stop until the problem is dealt with. An employer, the employees or **sub-contractors** could be taken to court.

The roles and responsibilities of the HSE are outlined in Table 1.3.

Responsibility	Explanation
Enforcement	It is the HSE's responsibility to reduce work-related death, injury and ill health. It will use the law against those who put others at risk.
Legislation and advice	The HSE will use health and safety legislation to serve improvement or prohibition notices or even to prosecute those who break health and safety rules. Inspectors will provide advice either face-to-face or in writing on health and safety matters.
Inspection	The HSE will look at site conditions, standards and practices and inspect documents to make sure that businesses and individuals are complying with health and safety law.

Table 1.3 HSE roles and responsibilities

Sources of health and safety information

There is a wide variety of health and safety information. Most of it is available free of charge, while other organisations may make a charge to provide information and advice. Table 1.4 outlines the key sources of health and safety information.

Source	Types of information	Website
Health and Safety Executive (HSE)	The HSE is the primary source of work-related health and safety information. It covers all possible topics and industries.	www.hse.gov.uk
Construction Industry Training Board (CITB)	The national training organisation provides key information on legislation and site safety.	www.citb.co.uk
British Standards Institute (BSI)	Provides guidelines for risk management, PPE, fire hazards and many other health and safety-related areas.	www.bsigroup.com
Royal Society for the Prevention of Accidents (RoSPA)	Provides training, consultancy and advice on a wide range of health and safety issues that are aimed to reduce work related accidents and ill health.	www.rospa.com
Royal Society for Public Health (RSPH)	Has a range of qualifications and training programmes focusing on health and safety.	www.rsph.org.uk

Table 1.4 Health and safety information

Informing the HSE

The HSE requires the reporting of:

* deaths and injuries – any **major injury**, **over 7-day injury** or death

* occupational disease

* dangerous occurrence – a collapse, explosion, fire or collision

* gas accidents – any accidental leaks or other incident related to gas.

Enforcing guidance

Work-related injuries and illnesses affect huge numbers of people. According to the HSE, 1.1 million working people in the UK suffered from a work-related illness in 2011 to 2012. Across all industries, 173 workers were killed, 111,000 other injuries were reported and 27 million working days were lost.

The construction industry is a high risk one and, although only around 5 per cent of the working population is in construction, it accounts for 10 per cent of all major injuries and 22 per cent of fatal injuries.

The good news is that enforcing guidance on health and safety has driven down the numbers of injuries and deaths in the industry. Only 20 years ago over 120 construction workers died in workplace accidents each year. This is now reduced to fewer than 60 a year.

However, there is still more work to be done and it is vital that organisations such as the HSE continue to enforce health and safety and continue to reduce risks in the industry.

On-site safety inductions and toolbox talks

The HSE suggests that all new workers arriving on site should attend a short induction session on health and safety. It should:

* show the commitment of the company to health and safety

* explain the health and safety policy

* explain the roles individuals play in the policy

* state that each individual has a legal duty to contribute to safe working

* cover issues like excavations, work at height, electricity and fire risk

* provide a layout of the site and show evacuation routes

* identify where fire fighting equipment is located

* ensure that all employees have evidence of their skills

* stress the importance of signing in and out of the site.

KEY TERMS

Major injury

– any fractures, amputations, dislocations, loss of sight or other severe injury.

Over 7-day injury

– an injury that has kept someone off work for more than seven days.

DID YOU KNOW?

Workplace injuries cost the UK £13.4bn in 2010 to 2011.

Behaviour and actions that could affect others

It is the responsibility of everyone on site not only to look after their own health and safety, but also to ensure that their actions do not put anyone else at risk.

Trying to carry out work that you are not competent to do is not only dangerous to yourself but could compromise the safety of others.

Simple actions, such as ensuring that all of your rubbish and waste is properly disposed of, will go a long way to removing hazards on site that could affect others.

Just as you should not create a hazard, ignoring an obvious one is just as dangerous. You should always obey site rules and particularly the health and safety rules. You should follow any instructions you are given.

ACCIDENT AND EMERGENCY PROCEDURES

All sites will have specific procedures for dealing with accidents and emergencies. An emergency will often mean that the site needs to be evacuated, so you should know in advance where to assemble and who to report to. The site should never be re-entered without authorisation from an individual in charge or the emergency services.

Types of emergencies

Emergencies are incidents that require immediate action. They can include:

* fires
* spillages or leaks of chemicals or other hazardous substances, such as gas
* failure of a scaffold
* collapse of a wall or trench
* a health problem
* an injury
* bombs and security alerts.

Legislation and reporting accidents

RIDDOR (1995) puts a duty on employers, anyone who is self-employed, or an individual in control of the work, to report any serious workplace accidents, occupational diseases or dangerous occurrences (also known as near misses).

The report has to be made by these individuals and, if it is serious enough, the responsible person may have to fill out a RIDDOR report.

Figure 1.2 It's important that you know where your company's fire-fighting equipment is located

Injuries, diseases and dangerous occurrences

Construction sites can be dangerous places, as we have seen. The HSE maintains a list of all possible injuries, diseases and dangerous occurrences, particularly those that need to be reported.

Injuries

There are two main classifications of injuries: minor and major. A minor injury can usually be handled by a competent first aider, although it is often a good idea to refer the individual to their doctor or to the hospital. Typical minor injuries can include:

* minor cuts
* minor burns
* exposure to fumes.

Major injuries are more dangerous and will usually require the presence of an ambulance with paramedics. Major injuries can include:

* bone fracture
* concussion
* unconsciousness
* electric shock.

Diseases

There are several different diseases and health issues that have to be reported, particularly if a doctor notifies that a disease has been diagnosed. These include:

* poisoning
* infections
* skin diseases
* occupational cancer
* lung diseases
* hand/arm vibration syndrome.

Dangerous occurrences

Even if something happens that does not result in an injury, but could easily have done so, it is classed as a dangerous occurrence. It needs to be reported immediately and then followed up by an accident report form. Dangerous occurrences can include:

* accidental release of a substance that could damage health

* anything coming into contact with overhead power lines

* an electrical problem that caused a fire or explosion

* collapse or partial collapse of scaffolding over 5 m high.

PRACTICAL TIP

An up-to-date list of dangerous occurrences is maintained by the Health and Safety Executive.

Recording accidents and emergencies

The Reporting of Injuries, Diseases and Dangerous Occurrences Regulations (RIDDOR) (1995) requires employers to:

* report any relevant injuries, diseases or dangerous occurrences to the Health and Safety Executive (HSE)

* keep records of incidents in a formal and organised manner (for example, in an accident book or online database).

After an accident, you may need to complete an accident report form – either in writing or online. This form may be completed by the person who was injured or the first aider.

On the accident report form you need to note down:

* the casualty's personal details, e.g. name, address, occupation

* the name of the person filling in the report form

* the details of the accident.

In addition, the person reporting the accident will need to sign the form.

On site a trained first aider will be the first individual to try and deal with the situation. In addition to trying to save life, stop the condition from getting worse and getting help, they will also record the occurrence.

On larger sites there will be a health and safety officer, who would keep records and documentation detailing any accidents and emergencies that have taken place on site. All companies should keep such records; it may be a legal requirement for them to do so under RIDDOR and it is good practice to do so in case the HSE asks to see it.

Importance of reporting accidents and near misses

Reporting incidents is not just about complying with the law or providing information for statistics. Each time an accident or near miss takes place it means lessons can be learned and future problems avoided.

The accident or near miss can alert the business or organisation to a potential problem. They can then take steps to ensure that it does not occur in the future.

Major and minor injuries and near misses

RIDDOR defines a major injury as:

* a fracture (but not to a finger, thumb or toes)

* a dislocation

* an amputation

* a loss of sight in an eye

* a chemical or hot metal burn to the eye

* a penetrating injury to the eye

* an electric shock or electric burn leading to unconsciousness and/or requiring resuscitation

* hyperthermia, heat-induced illness or unconsciousness

* asphyxia

* exposure to a harmful substance

* inhalation of a substance

* acute illness after exposure to toxins or infected materials.

A minor injury could be considered as any occurrence that does not fall into any of the above categories.

A near miss is any incident that did not actually result in an injury but which could have caused a major injury if it had done so. Non-reportable near misses are useful to record as they can help to identify potential problems. Looking at a list of near misses might show patterns for potential risk.

Accident trends

We have already seen that the HSE maintains statistics on the number and types of construction accidents. The following are among the 2011/2012 construction statistics:

* There were 49 fatalities.

* There were 5,000 occupational cancer patients.

* There were 74,000 cases of work-related ill health.

* The most common types of injury were caused by falls, although many injuries were caused by falling objects, collapses and electricity. A number of construction workers were also hurt when they slipped or tripped, or were injured while lifting heavy objects.

Accidents, emergencies and the employer

Even less serious accidents and injuries can cost a business a great deal of money. But there are other costs too:

* Poor company image – if a business does not have health and safety controls in place then it may get a reputation for not caring about its employees. The number of accidents and injuries may be far higher than average.

* Loss of production – the injured individual might have to be treated and then may need a period of time off work to recover. The loss of production can include those who have to take time out from working to help the injured person and the time of a manager or supervisor who has to deal with all the paperwork and problems.

* Insurance – each time there is an accident or injury claim against the company's insurance the premiums will go up. If there are many accidents and injuries the business may find it impossible to get insurance. It is a legal requirement for a business to have insurance so in the end that company might have to close down.

* Closure of site – if there is a serious accident or injury then the site may have to be closed while investigations take place to discover the reason, or who was responsible. This could cause serious delays and loss of income for workers and the business.

DID YOU KNOW?

RoSPA (the Royal Society for the Prevention of Accidents) uses many of the statistics from the HSE. The latest figures that RoSPA has analysed date back to 2008/2009. In that year, 1.2 million people in the UK were suffering from work-related illnesses. With fewer than 132,000 reportable injuries at work, this is believed to be around half of the real figure.

DID YOU KNOW?

An employee working in a small business broke two bones in his arm. He could not return to proper duties for eight months. He lost out on wages while he was off sick and, in total, it cost the business over £45,000.

REED TIP

On some construction sites, you may get a Health and Safety Inspector come to look round without any notice – one more reason to always be thinking about working safely.

Accident and emergency authorised personnel

Several different groups of people could be involved in dealing with accident and emergency situations. These are listed in Table 1.5.

Authorised personnel	Role
First aiders and emergency responders	These are employees on site and in the workforce who have been trained to be the first to respond to accidents and injuries. The minimum provision of an appointed person would be someone who has had basic first aid training. The appointment of a first aider is someone who has attained a higher or specific level of training. A construction site with fewer than 5 employees needs an appointed first aider. A construction site with up to 50 employees requires a trained first aider, and for bigger sites at least one trained first aider is required for every 50 people.
Supervisors and managers	These have the responsibility of managing the site and would have to organise the response and contact emergency services if necessary. They would also ensure that records of any accidents are completed and up to date and notify the HSE if required.
Health and Safety Executive	The HSE requires businesses to investigate all accidents and emergencies. The HSE may send an inspector, or even a team, to investigate and take action if the law has been broken.
Emergency services	Calling the emergency services depends on the seriousness of the accident. Paramedics will take charge of the situation if there is a serious injury and if they feel it necessary will take the individual to hospital.

Table 1.5 People who deal with accident and emergency situations

Figure 1.3 A typical first aid box

The basic first aid kit

BS 8599 relates to first aid kits, but it is not legally binding. The contents of a first aid box will depend on an employer's assessment of their likely needs. The HSE does not have to approve the contents of a first aid box but it states that where the work involves low level hazards the minimum contents of a first aid box should be:

* a copy of its leaflet on first aid – *HSE Basic advice on first aid at work*

* 20 sterile plasters of assorted size

* 2 sterile eye pads

* 4 sterile triangular bandages

* 6 safety pins

* 2 large sterile, unmedicated wound dressings

* 6 medium-sized sterile unmedicated wound dressings

* 1 pair of disposable gloves.

The HSE also recommends that no tablets or medicines are kept in the first aid box.

What to do if you discover an accident

When an accident happens it may not only injure the person involved directly, but it may also create a hazard that could then injure others. You need to make sure that the area is safe enough for you or someone else to help the injured person. It may be necessary to turn off the electrical supply or remove obstructions to the site of the accident.

The first thing that needs to be done if there is an accident is to raise the alarm. This could mean:

* calling for the first aider

* phoning for the emergency services

* dealing with the problem yourself.

How you respond will depend on the severity of the injury.

You should follow this procedure if you need to contact the emergency services:

* Find a telephone away from the emergency.

* Dial 999.

* You may have to go through a switchboard. Carefully listen to what the operator is saying to you and try to stay calm.

* When asked, give the operator your name and location, and the name of the emergency service or services you require.

* You will then be transferred to the appropriate emergency service, who will ask you questions about the accident and its location. Answer the questions in a clear and calm way.

* Once the call is over, make sure someone is available to help direct the emergency services to the location of the accident.

IDENTIFYING HAZARDS

As we have already seen, construction sites are potentially dangerous places. The most effective way of handling health and safety on a construction site is to spot the hazards and deal with them before they can cause an accident or an injury. This begins with basic housekeeping and carrying out risk assessments. It also means having a procedure in place to report hazards so that they can be dealt with.

Good housekeeping

Work areas should always be clean and tidy. Sites that are messy, strewn with materials, equipment, wires and other hazards can prove to be very dangerous. You should:

* always work in a tidy way

* never block fire exits or emergency escape routes

* never leave nails and screws scattered around

* ensure you clean and sweep up at the end of each working day

* not block walkways

* never overfill skips or bins

* never leave food waste on site.

Risk assessments and method statements

It is a legal requirement for employers to carry out risk assessments. This covers not only those who are actually working on a particular job, but other workers in the immediate area, and others who might be affected by the work.

It is important to remember that when you are carrying out work your actions may affect the safety of other people. It is important, therefore, to know whether there are any potential hazards. Once you know what these hazards are you can do something to either prevent or reduce them as a risk. Every job has potential hazards.

There are five simple steps to carrying out a risk assessment, which are shown in Table 1.6, using the example of repointing brickwork on the front face of a dwelling.

Step	Action	Example
1	Identify hazards	The property is on a street with a narrow pavement. The damaged brickwork and loose mortar need to be removed and placed in a skip below. Scaffolding has been erected. The road is not closed to traffic.
2	Identify who is at risk	The workers repointing are at risk as they are working at height. Pedestrians and vehicles passing are at risk from the positioning of the skip and the chance that debris could fall from height.
3	What is the risk from the hazard that may cause an accident?	The risk to the workers is relatively low as they have PPE and the scaffolding has been correctly erected. The risk to those passing by is higher, as they are unaware of the work being carried out above them.
4	Measures to be taken to reduce the risk	Station someone near the skip to direct pedestrians and vehicles away from the skip while the work is being carried out. Fix a secure barrier to the edge of the scaffolding to reduce the chance of debris falling down. Lower the bricks and mortar debris using a bucket or bag into the skip and not throwing them from the scaffolding. Consider carrying out the work when there are fewer pedestrians and less traffic on the road.
5	Monitor the risk	If there are problems with the first stages of the job, you need to take steps to solve them. If necessary consider taking the debris by hand through the building after removal.

Table 1.6 A five-step risk assessment for repointing brickwork

Your employer should follow these working practices, which can help to prevent accidents or dangerous situations occurring in the workplace:

* *Risk assessments* look carefully at what could cause an individual harm and how to prevent this. This is to ensure that no one should be injured or become ill as a result of their work. Risk assessments identify how likely it is that an accident might happen and the consequences of it happening. A risk factor is worked out and control measures created to try to offset them.

* *Method statements,* however brief, should be available for every risk assessment. They summarise risk assessments and other findings to provide guidance on how the work should be carried out.

* *Permit to work systems* are used for very high risk or even potentially fatal activities. They are checklists that need to be completed before the work begins. They must be signed by a supervisor.

* *A hazard book* lists standard tasks and identifies common hazards. These are useful tools to help quickly identify hazards related to particular tasks.

Types of hazards

Typical construction accidents can include:

* fires and explosions

* slips, trips and falls.

* burns, including those from chemicals

* falls from scaffolding, ladders and roofs

* electrocution

* injury from faulty machinery

* power tool accidents

* being hit by construction debris

* falling through holes in flooring

We will look at some of the more common hazards in a little more detail.

Fires
Fires need oxygen, heat and fuel to burn. Even a spark can provide enough heat needed to start a fire, and anything flammable, such as petrol, paper or wood, provides the fuel. It may help to remember the 'triangle of fire' – heat, oxygen and fuel are all needed to make fire so remove one or more to help prevent or stop the fire.

Tripping

Leaving equipment and materials lying around can cause accidents, as can trailing cables and spilt water or oil. Some of these materials are also potential fire hazards.

Chemical spills

If the chemicals are not hazardous then they just need to be mopped up. But sometimes they do involve hazardous materials and there will be an existing plan on how to deal with them. A risk assessment will have been carried out.

Falls from height

A fall even from a low height can cause serious injuries. Precautions need to be taken when working at height to avoid permanent injury. You should also consider falls into open excavations as falls from height. All the same precautions need to be in place to prevent a fall.

Burns

Burns can be caused not only by fires and heat, but also from chemicals and solvents. Electricity and wet concrete and cement can also burn skin. PPE is often the best way to avoid these dangers. Sunburn is a common and uncomfortable form of burning and sunscreen should be made available. For example, keeping skin covered up will help to prevent sunburn. You might think a tan looks good, but it could lead to skin cancer.

Electrical

Electricity is hazardous and electric shocks can cause burns and muscle damage, and can kill.

Exposure to hazardous substances

We look at hazardous substances in more detail on pages 20–1. COSHH regulations identify hazardous substances and require them to be labelled. You should always follow the instructions when using them.

Plant and vehicles

On busy sites there is always a danger from moving vehicles and heavy plant. Although many are fitted with reversing alarms, it may not be easy to hear them over other machinery and equipment. You should always ensure you are not blocking routes or exits. Designated walkways separate site traffic and pedestrians – this includes workers who are walking around the site. Crossing points should be in place for ease of movement on site.

Reporting hazards

We have already seen that hazards have the potential to cause serious accidents and injuries. It is therefore important to report hazards and there are different methods of doing this.

The first major reason to report hazards is to prevent danger to others, whether they are other employees or visitors to the site. It is vital to prevent accidents from taking place and to quickly correct any dangerous situations.

Injuries, diseases and actual accidents all need to be reported and so do dangerous occurrences. These are incidents that do not result in an actual injury, but could easily have hurt someone.

Accidents need to be recorded in an accident book, computer database or other secure recording system, as do near misses. Again it is a legal requirement to keep appropriate records of accidents and every company will have a procedure for this which they should tell you about. Everyone should know where the book is kept or how the records are made. Anyone that has been hurt or has taken part in dealing with an occurrence should complete the details of what has happened. Typically this will require you to fill in:

* the date, time and place of the incident

* how it happened

* what was the cause

* how it was dealt with

* who was involved

* signature and date.

The details in the book have to be transferred onto an official HSE report form.

As far as is possible, the site, company or workplace will have set procedures in place for reporting hazards and accidents. These procedures will usually be found in the place where the accident book or records are stored. The location tends to be posted on the site notice board.

How hazards are created

Construction sites are busy places. There are constantly new stages in development. As each stage is begun a whole new set of potential hazards need to be considered.

At the same time, new workers will always be joining the site. It is mandatory for them to be given health and safety instruction during induction. But sometimes this is impossible due to pressure of work or availability of trainers.

Construction sites can become even more hazardous in times of extreme weather:

* Flooding – long periods of rain can cause trenches to fill with water, cellars to be flooded and smooth surfaces to become extremely wet and slippery.

* Wind – strong winds may prevent all work at height. Scaffolding may have become unstable, unsecured roofing materials may come loose, dry-stored materials such as sand and cement may have been blown across the site.

* Heat – this can change the behaviour of materials: setting quicker, failing to cure and melting. It can also seriously affect the health of the workforce through dehydration and heat exhaustion.

* Snow – this can add enormous weight to roofs and other structures and could cause collapse. Snow can also prevent access or block exits and can mean that simple and routine work becomes impossible due to frozen conditions.

Storing combustibles and chemicals

A combustible substance can be both flammable and explosive. There are some basic suggestions from the HSE about storing these:

* Ventilation – the area should be well ventilated to disperse any vapours that could trigger off an explosion.

* Ignition – an ignition is any spark or flame that could trigger off the vapours, so materials should be stored away from any area that uses electrical equipment or any tool that heats up.

* Containment – the materials should always be kept in proper containers with lids and there should be spillage trays to prevent any leak seeping into other parts of the site.

* Exchange – in many cases it can be possible to find an alternative material that is less dangerous. This option should be taken if possible.

* Separation – always keep flammable substances away from general work areas. If possible they should be partitioned off.

Combustible materials can include a large number of commonly used substances, such as cleaning agents, paints and adhesives.

HEALTH AND HYGIENE

Just as hazards can be a major problem on site, other less obvious problems relating to health and hygiene can also be an issue. It is both your responsibility and that of your employer to make sure that you stay healthy.

The employer will need to provide basic welfare facilities, no matter where you are working and these must have minimum standards.

KEY TERMS

Contamination

– this is when a substance has been polluted by some harmful substance or chemical.

Welfare facilities

Welfare facilities can include a wide range of different considerations, as can be seen in Table 1.7

Facilities	Purpose and minimum standards
Toilets	If there is a lock on the door there is no need to have separate male and female toilets. There should be enough for the site workforce. If there is no flushing water on site they must be chemical toilets.
Washing facilities	There should be a wash basin large enough to be able to wash up to the elbow. There should be soap, hot and cold water and, if you are working with dangerous substances, then showers are needed.
Drinking water	Clean drinking water should be available; either directly connected to the mains or bottled water. Employers must ensure that there is no contamination.
Dry room	This can operate also as a store room, which needs to be secure so that workers can leave their belongings there and also use it as a place to dry out if they have been working in wet weather, in which case a heater needs to be provided.
Work break area	This is a shelter out of the wind and rain, with a kettle, a microwave, tables and chairs. It should also have heating.

Table 1.7 Welfare facilities in the workplace

CASE STUDY

South Tyneside Homes

South Tyneside Council's Housing Company

Staying safe on site

Johnny McErlane finished his apprenticeship at South Tyneside Homes a year ago.

'I've been working on sheltered accommodation for the last year, so there are a lot of vulnerable and elderly people around. All the things I learnt at college from doing the health and safety exams comes into practice really, like taking care when using extension leads, wearing high-vis and correct footwear. It's not just about your health and safety, but looking out for others as well.

On the shelters, you can get a health and safety inspector who just comes around randomly, so you have to always be ready. It just becomes a habit once it's been drilled into you. You're health and safety conscious all the time.

The shelters also have a fire alarm drill every second Monday, so you've got to know the procedure involved there. When it comes to the more specialised skills, such as mouth-to-mouth and CPR, you might have a designated first aider on site who will have their skills refreshed regularly. Having a full first aid certificate would be valuable if you're working in construction.

You cover quite a bit of the first aid skills in college and you really have to know them because you're not always working on large sites. For example, you might be on the repairs team, working in people's houses where you wouldn't have a first aider, so you've got to have the basic knowledge yourself, just in case. All our vans have a basic first aid kit that's kept fully stocked.

The company keeps our knowledge current with these "toolbox talks", which are like refresher courses. They give you any new information that needs to be passed on to all the trades. It's a good way of keeping everyone up to date.'

Noise

Ear defenders are the best precaution to protect the ears from loud noises on site. Ear defenders are either basic ear plugs or ear muffs, which can be seen in Fig 1.13 on page 32.

The long-term impact of noise depends on the intensity and duration of the noise. Basically, the louder and longer the noise exposure, the more damage is caused. There are ways of dealing with this:

* Remove the source of the noise.

* Move the equipment away from those not directly working with it.

* Put the source of the noise into a soundproof area or cover it with soundproof material.

* Ask a supervisor if they can move all other employees away from that part of the site until the noise stops.

Substances hazardous to health

COSHH Regulations (see page 3) identify a wide variety of substances and materials that must be labelled in different ways.

Controlling the use of these substances is always difficult. Ideally, their use should be eliminated (stopped) or they should be replaced with something less harmful. Failing this, they should only be used in controlled or restricted areas. If none of this is possible then they should only be used in controlled situations.

If a hazardous situation occurs at work, then you should:

* ensure the area is made safe

* inform the supervisor, site manager, safety officer or other nominated person.

You will also need to report any potential hazards or near misses.

Personal hygiene

Construction sites can be dirty places to work. Some jobs will expose you to dust, chemicals or substances that can make contact with your skin or may stain your work clothing. It is good practice to wear suitable PPE as a first line of defence as chemicals can penetrate your skin. Whenever you have finished a job you should always wash your hands. This is certainly true before eating lunch or travelling home. It can be good practice to have dedicated work clothing, which should be washed regularly.

Always ensure you wash your hands and face and scrub your nails. This will prevent dirt, chemicals and other substances from contaminating your food and your home.

Make sure that you regularly wash your work clothing and either repair it or replace it if it becomes too worn or stained.

Health risks

The construction industry uses a wide variety of substances that could harm your health. You will also be carrying out work that could be a health risk to you, and you should always be aware that certain activities could cause long-term damage or even kill you if things go wrong. Unfortunately not all health risks are immediately obvious. It is important to make sure that from time to time you have health checks, particularly if you have been using hazardous substances. Table 1.8 outlines some potential health risks in a typical construction site.

KEY TERMS

Dermatitis

– this is an inflammation of the skin. The skin will become red and sore, particularly if you scratch the area. A GP should be consulted.

Leptospirosis

– this is also known as Weil's disease. It is spread by touching soil or water contaminated with the urine of wild animals infected with the leptospira bacteria. Symptoms are usually flu-like but in extreme cases it can cause organ failure.

Health risk	Potential future problems
Dust	The most dangerous potential dust is, of course, asbestos, which **should only be handled by specialists under controlled conditions**. But even brick dust and other fine particles can cause eye injuries, problems with breathing and even cancer.
Chemicals	Inhaling or swallowing dangerous chemicals could cause immediate, long-term damage to lungs and other internal organs. Skin problems include burns or skin can become very inflamed and sore. This is known as dermatitis.
Bacteria	Contact with waste water or soil could lead to a bacterial infection. The germs in the water or dirt could cause infection which will require treatment if they enter the body. The most extreme version is leptospirosis.
Heavy objects	Lifting heavy, bulky or awkward objects can lead to permanent back injuries that could require surgery. Heavy objects can also damage the muscles in all areas of the body.
Noise	Failure to wear ear defenders when you are exposed to loud noises can permanently affect your hearing. This could lead to deafness in the future.
Vibrating tools	Using machines that vibrate can cause a condition known as hand/arm vibration syndrome (HAVS) or vibration white finger, which is caused by injury to nerves and blood vessels. You will feel tingling that could lead to permanent numbness in the fingers and hands, as well as muscle weakness.
Cuts	Any open wound, no matter how small, leaves your body exposed to potential infections. Cuts should always be cleaned and covered, preferably with a waterproof dressing. The blood loss from deep cuts could make you feel faint and weak, which may be dangerous if you are working at height or operating machinery.
Sunlight	Most construction work involves working outside. There is a temptation to take advantage of hot weather and get a tan. But long-term exposure to sunshine means risking skin cancer so you should cover up and apply sun cream.
Head injuries	You should seek medical attention after any bump to the head. Severe head injuries could cause epilepsy, hearing problems, brain damage or death.

Table 1.8 Health risks in construction

HANDLING AND STORING MATERIALS AND EQUIPMENT

On a busy construction site it is often tempting not to even think about the potential dangers of handling equipment and materials. If something needs to be moved or collected you will just pick it up without any thought. It is also tempting just to drop your tools and other equipment when you have finished with them to deal with later. But abandoned equipment and tools can cause hazards both for you and for other people.

Safe lifting

Lifting or handling heavy or bulky items is a major cause of injuries on construction sites. So whenever you are dealing with a heavy load, it is important to carry out a basic risk assessment.

The first thing you need to do is to think about the job to be done and ask:

* Do I need to lift it manually or is there another way of getting the object to where I need it?

Consider any mechanical methods of transporting loads or picking up materials. If there really is no alternative, then ask yourself:

1. Do I need to bend or twist?
2. Does the object need to be lifted or put down from high up?
3. Does the object need to be carried a long way?
4. Does the object need to be pushed or pulled for a long distance?
5. Is the object likely to shift around while it is being moved?

If the answer to any of these questions is 'yes', you may need to adjust the way the task is done to make it safer.

Think about the object itself. Ask:

1. Is it just heavy or is it also bulky and an awkward shape?
2. How easy is it to get a good hand-hold on the object?
3. Is the object a single item or are there parts that might move around and shift the weight?
4. Is the object hot or does it have sharp edges?

Again, if you have answered 'yes' to any of these questions, then you need to take steps to address these issues.

It is also important to think about the working environment and where the lifting and carrying is taking place. Ask yourself:

1. Are the floors stable?
2. Are the surfaces slippery?
3. Will a lack of space restrict my movement?
4. Are there any steps or slopes?
5. What is the lighting like?

Before lifting and moving an object, think about the following:

* Check that your pathway is clear to where the load needs to be taken.

* Look at the product data sheet and assess the weight. If you think the object is too heavy or difficult to move then ask someone to help you. Alternatively, you may need to use a mechanical lifting device.

When you are ready to lift, gently raise the load. Take care to ensure the correct posture – you should have a straight back, with your elbows tucked in, your knees bent and your feet slightly apart.

Once you have picked up the load, move slowly towards your destination. When you get there, make sure that you do not drop the load but carefully place it down.

> **DID YOU KNOW?**
>
> Although many people regard the weight limit for lifting and/or moving heavy or awkward objects to be 20 kg, the HSE does not recommend safe weights. There are many things that will affect the ability of an individual to lift and carry particular objects and the risk that this creates, so manual handling should be avoided altogether where possible.

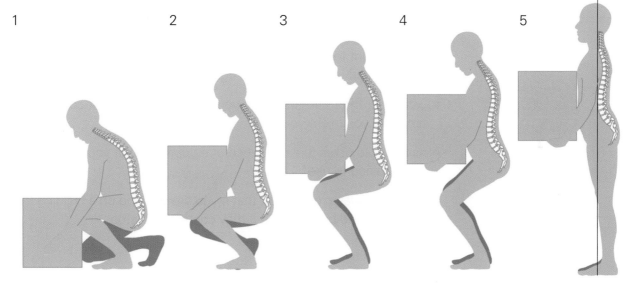

Figure 1.4 Take care to follow the correct procedure for lifting

Sack trolleys are useful for moving heavy and bulky items around. Gently slide the bottom of the sack trolley under the object and then raise the trolley to an angle of 45° before moving off. Make sure that the object is properly balanced and is not too big for the trolley.

Trailers and forklift trucks are often used on large construction sites, as are dump trucks. Never use these without proper training.

Figure 1.5 Pallet truck

Figure 1.6 Sack trolley

Site safety equipment

You should always read the construction site safety rules and when required wear your PPE. Simple things, such as wearing the right footwear for the right job, are important.

Safety equipment falls into two main categories:

* PPE – including hard hats, footwear, gloves, glasses and safety vests

* perimeter safety – this includes screens, netting and guards or clamps to prevent materials from falling or spreading.

Construction safety is also directed by signs, which will highlight potential hazards.

Safe handling of materials and equipment

All tools and equipment are potentially dangerous. It is up to you to make sure that they do not cause harm to yourself or others. You should always know how to use tools and equipment. This means either instruction from someone else who is experienced, or at least reading the manufacturer's instructions.

You should always make sure that you:

* use the right tool – don't be tempted to use a tool that is close to hand instead of the one that is right for the job

* wear your PPE – the one time you decide not to bother could be the time that you injure yourself

* never try to use a tool or a piece of equipment that you have not been trained to use.

You should always remember that if you are working on a building that was constructed before 2000 it may contain asbestos.

Correct storage

We have already seen that tools and equipment need to be treated with respect. Damaged tools and equipment are not only less effective at doing their job, they could also cause you to injure yourself.

Table 1.9 provides some pointers on how to store and handle different types of materials and equipment.

Materials and equipment	Safe storage and handling
Hand tools	Store hand tools with sharp edges either in a cover or a roll. They should be stored in bags or boxes. They should always be dried before putting them away as they will rust.
Power tools	Never carry them by the cable. Store them in their original carrying case. Always follow the manufacturer's instructions.
Wheelbarrows	Check the tyres and metal stays regularly. Always clean out after use and never overload.
Bricks and blocks	Never store more than two packs high. When cutting open a pack, be careful as the bricks could collapse.
Slabs and curbs	Store slabs flat on their edges on level ground, preferably with wood underneath to prevent damage. Store curbs the same way. To prevent weather damage, cover them with a sheet.
Tiles	Always cover them and protect them from damage as they are relatively fragile. Ideally store them in a hut or container.
Aggregates	Never store aggregates under trees as leaves will drop on them and contaminate them. Cover them with plastic sheets.
Plaster and plasterboard	Plaster needs to be kept dry, so even if stored inside you should take the precaution of putting the bags on pallets. To prevent moisture do not store against walls and do not pile higher than five bags. Plasterboard can be awkward to manage and move around. It also needs to be stored in a waterproof area. It should be stored flat and off the ground but should not be stored against walls as it may bend. Use a rotation system so that the materials are not stored in the same place for long periods.
Wood	Always keep wood in dry, well-ventilated conditions. If it needs to be stored outside it should be stored on bearers that may be on concrete. If wood gets wet and bends it is virtually useless. Always be careful when moving large cuts of wood or sheets of ply or MDF as they can easily become damaged.
Adhesives and paint	Always read the manufacturer's instructions. Ideally they should always be stored on clearly marked shelves. Make sure you rotate the stock using the older stock first. Always make sure that containers are tightly sealed. Storage areas must comply with fire regulations and display signs to advise of their contents.

Table 1.9 Safe storing and handling of materials and equipment

Waste control

The expectation within the building services industry is increasingly that working practices conserve energy and protect the environment. Everyone can play a part in this. For example, you can contribute by turning off hose pipes when you have finished using water, or not running electrical items when you don't need to.

Simple things, such as keeping construction sites neat and orderly, can go a long way to conserving energy and protecting the environment. A good way to remember this is Sort, Set, Shine, Standardise:

* Sort – sort and store items in your work area, eliminate clutter and manage deliveries.

* Set – everything should have its own place and be clearly marked and easy to access. In other words, be neat!

Figure 1.7 It's important to create as little waste as possible on the construction site

* Shine – clean your work area and you will be able to see potential problems far more easily.

* Standardise – by using standardised working practices you can keep organised, clean and safe.

Reducing waste is all about good working practice. By reducing wastage disposal, and recycling materials on site, you will benefit from savings on raw materials and lower transportation costs.

Planning ahead, and accurately measuring and cutting materials, means that you will be able to reduce wastage.

BASIC WORKING PLATFORMS AND ACCESS EQUIPMENT

Working at height should be eliminated or the work carried out using other methods where possible. However, there may be situations where you may need to work at height. These situations can include:

* roofing

* repair and maintenance above ground level

* working on high ceilings.

Any work at height must be carefully planned. Access equipment includes all types of ladder, scaffold and platform. You must always use a working platform that is safe. Sometimes a simple step ladder will be sufficient, but at other times you may have to use a tower scaffold.

Generally, ladders are fine for small, quick jobs of less than 30 minutes. However, for larger, longer jobs a more permanent piece of access equipment will be necessary.

Working platforms and access equipment: good practice and dangers of working at height

Table 1.10 outlines the common types of equipment used to allow you to work at heights, along with the basic safety checks necessary.

Equipment	Main features	Safety checks
Step ladder	Ideal for confined spaces. Four legs give stability	• Knee should remain below top of steps • Check hinges, cords or ropes • Position only to face work
Ladder	Ideal for basic access, short-term work. Made from aluminium, fibreglass or wood	• Check rungs, tie rods, repairs, and ropes and cords on stepladders • Ensure it is placed on firm, level ground • Angle should be no greater than 75° or 1 in 4
Mobile mini towers or scaffolds	These are usually aluminium and foldable, with lockable wheels	• Ensure the ground is even and the wheels are locked • Never move the platform while it has tools, equipment or people on it
Roof ladders and crawling boards	The roof ladder allows access while crawling boards provide a safe passage over tiles	• The ladder needs to be long enough and supported • Check boards are in good condition • Check the welds are intact • Ensure all clips function correctly
Mobile tower scaffolds	These larger versions of mini towers usually have edge protection	• Ensure the ground is even and the wheels are locked • Never move the platform while it has tools, equipment or people on it • Base width to height ratio should be no greater than 1:3
Fixed scaffolds and edge protection	Scaffolds fitted and sized to the specific job, with edge protection and guard rails	• There needs to be sufficient braces, guard rails and scaffold boards • The tubes should be level • There should be proper access using a ladder
Mobile elevated work platforms	Known as scissor lifts or cherry pickers	• Specialist training is required before use • Use guard rails and toe boards • Care needs to be taken to avoid overhead hazards such as cables

Table 1.10 Equipment for working at height and safety checks

You must be trained in the use of certain types of access equipment, like mobile scaffolds. Care needs to be taken when assembling and using access equipment. These are all examples of good practice:

* Step ladders should always rest firmly on the ground. Only use the top step if the ladder is part of a platform.

* Do not rest ladders against fragile surfaces, and always use both hands to climb. It is best if the ladder is steadied (footed) by someone at the foot of the ladder. Always maintain three points of contact – two feet and one hand.

* A roof ladder is positioned by turning it on its wheels and pushing it up the roof. It then hooks over the ridge tiles. Ensure that the access ladder to the roof is directly beside the roof ladder.

* A mobile scaffold is put together by slotting sections until the required height is reached. The working platform needs to have a suitable edge protection such as guard-rails and toe-boards. Always push from the bottom of the base and not from the top to move it, otherwise it may lean or topple over.

Figure 1.8 A tower scaffold

WORKING SAFELY WITH ELECTRICITY

It is essential whenever you work with electricity that you are competent and that you understand the common dangers. Electrical tools must be used in a safe manner on site. There are precautions that you can take to prevent possible injury, or even death.

Precautions

Whether you are using electrical tools or equipment on site, you should always remember the following:

* Use the right tool for the job.

* Use a transformer with equipment that runs on 110V.

* Keep the two voltages separate from each other. You should avoid using 230V where possible but, if you must, use a residual current device (RCD) if you have to use 230V.

* When using 110V, ensure that leads are yellow in colour.

* Check the plug is in good order

* Confirm that the fuse is the correct rating for the equipment.

* Check the cable (including making sure that it does not present a tripping hazard).

* Find out where the mains switch is, in case you need to turn off the power in the event of an emergency.

* Never attempt to repair electrical equipment yourself.

* Disconnect from the mains power before making adjustments, such as changing a drill bit.

* Make sure that the electrical equipment has a sticker that displays a recent test date.

Visual inspection and testing is a three-stage process:

1. The user should check for potential danger signs, such as a frayed cable or cracked plug.

2. A formal visual inspection should then take place. If this is done correctly then most faults can be detected.

3. Combined inspections and **PAT** should take place at regular intervals by a competent person.

Watch out for the following causes of accidents – they would also fail a safety check:

KEY TERMS

PAT

– Portable Appliance Testing – regular testing is a health and safety requirement under the Electricity at Work Regulations (1989).

* damage to the power cable or plug

* taped joints on the cable

* wet or rusty tools and equipment

* weak external casing

* loose parts or screws

* signs of overheating

* the incorrect fuse

* lack of cord grip

* electrical wires attached to incorrect terminals

* bare wires.

When preparing to work on an electrical circuit, do not start until a permit to work has been issued by a supervisor or manager to a competent person.

Make sure the circuit is broken before you begin. A 'dead' circuit will not cause you, or anybody else, harm. These steps must be followed:

* Switch off – ensure the supply to the circuit is switched off by disconnecting the supply cables or using an isolating switch.

* Isolate – disconnect the power cables or use an isolating switch.

* Warn others – to avoid someone reconnecting the circuit, place warning signs at the isolation point.

* Lock off – this step physically prevents others from reconnecting the circuit.

* Testing – is carried out by electricians but you should be aware that it involves three parts:

  1. testing a voltmeter on a known good source (a live circuit) so you know it is working properly

  2. checking that the circuit to be worked on is dead

  3. rechecking your voltmeter on the known live source, to prove that it is still working properly.

It is important to make sure that the correct point of isolation is identified. Isolation can be next to a local isolation device, such as a plug or socket, or a circuit breaker or fuse.

The isolation should be locked off using a unique key or combination. This will prevent access to a main isolator until the work has been completed. Alternatively, the handle can be made detachable in the OFF position so that it can be physically removed once the circuit is switched off.

Dangers

You are likely to encounter a number of potential dangers when working with electricity on construction sites or in private houses. Table 1.11 outlines the most common dangers.

DID YOU KNOW?

All power tools should be checked by the user before use. A PAT programme of maintenance, inspection and testing is necessary. The frequency of inspection and testing will depend on the appliance. Equipment is usually used for a maximum of three months between tests.

Danger	Identifying the danger
Faulty electrical equipment	Visually inspect for signs of damage. Equipment should be double insulated or incorporate an earth cable.
Damaged or worn cables	Check for signs of wear or damage regularly. This includes checking power tools and any wiring in the property.
Trailing cables	Cables lying on the ground, or worse, stretched too far, can present a tripping hazard. They could also be cut or damaged easily.
Cables and pipe work	Always treat services you find as though they are live. This is very important as services can be mistaken for one another. You may have been trained to use a cable and pipe locator that finds cables and metal pipes.
Buried or hidden cables	Make sure you have plans. Alternatively, use a cable and pipe locator, mark the positions, look out for signs of service connection cables or pipes and hand-dig trial holes to confirm positions.
Inadequate over-current protection	Check circuit breakers and fuses are the correct size current rating for the circuit. A qualified electrician may have to identify and label these.

Table 1.11 Common dangers when working with electricity

Each year there are around 1,000 accidents at work involving electric shocks or burns from electricity. If you are working in a construction site you are part of a group that is most at risk. Electrical accidents happen when you are working close to equipment that you think is disconnected but which is, in fact, live.

Another major danger is when electrical equipment is either misused or is faulty. Electricity can cause fires and contact with the live parts can give you an electric shock or burn you.

Different voltages

The two most common voltages that are used in the UK are 230V and 110V:

* 230V: this is the standard domestic voltage. But on construction sites it is considered to be unsafe and therefore 110V is commonly used.

* 110V: these plugs are marked with a yellow casement and they have a different shaped plug. A transformer is required to convert 230V to 110V.

Some larger homes, as well as industrial and commercial buildings, may have 415V supplies. This is the same voltage that is found on overhead electricity cables. In most houses and other buildings the voltage from these cables is reduced to 230V. This is what most electrical equipment works from. Some larger machinery actually needs 415V.

In these buildings the 415V comes into the building and then can either be used directly or it is reduced so that normal 230V appliances can be used.

Colour coded cables

Normally you will come across three differently coloured wires: Live, Neutral and Earth. These have standard colours that comply with European safety standards and to ensure that they are easily identifiable. However, in some older buildings the colours are different.

Wire type	Modern colour	Older colour
Live	Brown	Red
Neutral	Blue	Black
Earth	Yellow and Green	Yellow and Green

Table 1.12 Colour coding of cables

Working with equipment with different electrical voltages

You should always check that the electrical equipment that you are going to use is suitable for the available electrical supply. The equipment's power requirements are shown on its rating plate. The voltage from the supply needs to match the voltage that is required by the equipment.

Storing electrical equipment

Electrical equipment should be stored in dry and secure conditions. Electrical equipment should never get wet but – if it does happen – it should be dried before storage. You should always clean and adjust the equipment before connecting it to the electricity supply.

PERSONAL PROTECTIVE EQUIPMENT (PPE)

Personal protective equipment, or PPE, is a general term that is used to describe a variety of different types of clothing and equipment that aim to help protect against injuries or accidents. Some PPE you will use on a daily basis and others you may use from time to time. The type of PPE you wear depends on what you are doing and where you are. For example, the practical exercises in this book were photographed at a college, which has rules and requirements for PPE that are different to those on large construction sites. Follow your tutor's or employer's instructions at all times.

Types of PPE

PPE literally covers from head to foot. Here are the main PPE types.

Figure 1.9 A hi-vis jacket

Figure 1.10 Safety glasses and goggles

Figure 1.11 Hand protection

Figure 1.12 Head protection

Figure 1.13 Hearing protection

Protective clothing

Clothing protection such as overalls:

* provides some protection from spills, dust and irritants
* can help protect you from minor cuts and abrasions
* reduces wear to work clothing underneath.

Sometimes you may need waterproof or chemical-resistant overalls.

High visibility (hi-vis) clothing stands out against any background or in any weather conditions. It is important to wear high visibility clothing on a construction site to ensure that people can see you easily. In addition, workers should always try to wear light-coloured clothing underneath, as it is easier to see.

You need to keep your high visibility and protective clothing clean and in good condition.

Employers need to make sure that employees understand the reasons for wearing high visibility clothing and the consequences of not doing so.

Eye protection

For many jobs, it is essential to wear goggles or safety glasses to prevent small objects, such as dust, wood or metal, from getting into the eyes. As goggles tend to steam up, particularly if they are being worn with a mask, safety glasses can often be a good alternative.

Hand protection

Wearing gloves will help to prevent damage or injury to the hands or fingers. For example, general purpose gloves can prevent cuts, and rubber gloves can prevent skin irritation and inflammation, such as contact dermatitis caused by handling hazardous substances. There are many different types of gloves available, including specialist gloves for working with chemicals.

Head protection

Hard hats or safety helmets are compulsory on building sites. They can protect you from falling objects or banging your head. They need to fit well and they should be regularly inspected and checked for cracks. Worn straps mean that the helmet should be replaced, as a blow to the head can be fatal. Hard hats bear a date of manufacture and should be replaced after about 3 years.

Hearing protection

Ear defenders, such as ear protectors or plugs, aim to prevent damage to your hearing or hearing loss when you are working with loud tools or are involved in a very noisy job.

Respiratory protection

Breathing in fibre, dust or some gases could damage the lungs. Dust is a very common danger, so a dust mask, face mask or respirator may be necessary.

Make sure you have the right mask for the job. It needs to fit properly otherwise it will not give you sufficient protection.

Foot protection

Foot protection is compulsory on site, particularly if you are undertaking heavy work. Footwear should include steel toecaps (or equivalent) to protect feet against dropped objects, midsole protection (usually a steel plate) to protect against puncture or penetration from things like nails on the floor and soles with good grip to help prevent slips on wet surfaces.

Figure 1.14 Respiratory protection

Legislation covering PPE

The most important piece of legislation is the Personal Protective Equipment at Work Regulations (1992). It covers all sorts of PPE and sets out your responsibilities and those of the employer. Linked to this are the Control of Substances Hazardous to Health (2002) and the Provision and Use of Work Equipment Regulations (1992 and 1998).

Storing and maintaining PPE

All forms of PPE will be less effective if they are not properly maintained. This may mean examining the PPE and either replacing or cleaning it, or if relevant testing or repairing it. PPE needs to be stored properly so that it is not damaged, contaminated or lost. Each type of PPE should have a CE mark. This shows that it has met the necessary safety requirements.

Importance of PPE

PPE needs to be suitable for its intended use and it needs to be used in the correct way. As a worker or an employee you need to:

* make sure you are trained to use PPE

* follow your employer's instructions when using the PPE and always wear it when you are told to do so

* look after the PPE and if there is a problem with it report it.

Your employer will:

* know the risks that the PPE will either reduce or avoid

* know how the PPE should be maintained

* know its limitations.

Consequences of not using PPE

The consequences of not using PPE can be immediate or long-term. Immediate problems are more obvious, as you may injure yourself. The longer-term consequences could be ill health in the future. If your employer has provided PPE, you have a legal responsibility to wear it.

FIRE AND EMERGENCY PROCEDURES

KEY TERMS

Assembly point

– an agreed place outside the building to go to if there is an emergency.

If there is a fire or an emergency, it is vital that you raise the alarm quickly. You should leave the building or site and then head for the **assembly point.**

When there is an emergency a general alarm should sound. If you are working on a larger and more complex construction site, evacuation may begin by evacuating the area closest to the emergency. Areas will then be evacuated one-by-one to avoid congestion of the escape routes.

Three elements essential to creating a fire

Three ingredients are needed to make something combust (burn):

* oxygen   * heat   * fuel.

The fuel can be anything which burns, such as wood, paper or flammable liquids or gases, and oxygen is in the air around us, so all that is needed is sufficient heat to start a fire.

The fire triangle represents these three elements visually. By removing one of the three elements the fire can be prevented or extinguished.

Figure 1.15 Assembly point sign

How fire is spread

Fire can easily move from one area to another by finding more fuel. You need to consider this when you are storing or using materials on site, and be aware that untidiness can be a fire risk. For example, if there are wood shavings on the ground the fire can move across them, burning up the shavings.

Heat can also transfer from one source of fuel to another. If a piece of wood is on fire and is against or close to another piece of wood, that too will catch fire and the fire will have spread.

Figure 1.16 The fire triangle

On site, fires are classified according to the type of material that is on fire. This will determine the type of fire-fighting equipment you will need to use. The five different types of fire are shown in Table 1.13.

Class of fire	Fuel or material on fire
A	Wood, paper and textiles
B	Petrol, oil and other flammable liquids
C	LPG, propane and other flammable gases
D	Metals and metal powder
E	Electrical equipment

Table 1.13 Different classes of fire

There is also F, cooking oil, but this is less likely to be found on site, except in a kitchen.

Taking action if you discover a fire and fire evacuation procedures

During induction, you will have been shown what to do in the event of a fire and told about assembly points. These are marked by signs and somewhere on the site there will be a map showing their location.

If you discover a fire you should:

* sound the alarm

* not attempt to fight the fire unless you have had fire marshal training

* otherwise stop work, do not collect your belongings, do not run, and do not re-enter the site until the all clear has been given.

Different types of fire extinguishers

Extinguishers can be effective when tackling small localised fires. However, you must use the correct type of extinguisher. For example, putting water on an oil fire could make it explode. For this reason, you should not attempt to use a fire extinguisher unless you have had proper training.

When using an extinguisher it is important to remember the following safety points:

* Only use an extinguisher at the early stages of a fire, when it is small.

* The instructions for use appear on the extinguisher.

* If you do choose to fight the fire because it is small enough, and you are sure you know what is burning, position yourself between the fire and the exit, so that if it doesn't work you can still get out.

Type of fire risk	Fire class Symbol	White label Water	Cream label Foam	Black label Carbon dioxide	Blue label Dry powder	Yellow label Wet chemical
A – Solid (e.g. wood or paper)	A	✓	✓	✗	✓	✓
B – Liquid (e.g. petrol)	B	✗	✓	✓	✓	✗
C – Gas (e.g. propane)	C	✗	✗	✓	✓	✗
D – Metal (e.g. aluminium)	D METAL	✗	✗	✗	✓	✗
E – Electrical (i.e. any electrical equipment)	E	✗	✗	✓	✓	✗
F – Cooking oil (e.g. a chip pan)	F	✗	✗	✗	✗	✓

Table 1.14 Types of fire extinguishers

There are some differences you should be aware of when using different types of extinguisher:

* CO_2 *extinguishers* – do not touch the nozzle; simply operate by holding the handle. This is because the nozzle gets extremely cold when ejecting the CO_2, as does the canister. Fires put out with a CO_2 extinguisher may reignite, and you will need to ventilate the room after use.

* *Powder extinguishers* – these can be used on lots of kinds of fire, but can seriously reduce visibility by throwing powder into the air as well as on the fire.

SIGNS AND SAFETY NOTICES

In a well-organised working environment safety signs will warn you of potential dangers and tell you what to do to stay safe. They are used to warn you of hazards. Their purpose is to prevent accidents. Some will tell you what to do (or not to do) in particular parts of the site and some will show you where things are, such as the location of a first aid box or a fire exit.

Types of signs and safety notices

There are five basic types of safety sign, as well as signs that are a combination of two or more of these types. These are shown in Table 1.15.

Type of safety sign	What it tells you	What it looks like	Example
Prohibition sign	Tells you what you must *not* do	Usually round, in red and white	Do not use ladder
Hazard sign	Warns you about hazards	Triangular, in yellow and black	Caution Slippery floor
Mandatory sign	Tells you what you *must* do	Round, usually blue and white	Masks must be worn in this area
Safe condition or information sign	Gives important information, e.g. about where to find fire exits, assembly points or first aid kit, or about safe working practices	Green and white	First aid
Firefighting sign	Gives information about extinguishers, hydrants, hoses and fire alarm call points, etc.	Red with white lettering	Fire alarm call point
Combination sign	These have two or more of the elements of the other types of sign, e.g. hazard, prohibition and mandatory		DANGER Isolate before removing cover

Table 1.15 Different types of safety signs

TEST YOURSELF

1. Which of the following requires you to tell the HSE about any injuries or diseases?

   a. HASAWA

   b. COSHH

   c. RIDDOR

   d. PUWER

2. What is a prohibition notice?

   a. An instruction from the HSE to stop all work until a problem is dealt with

   b. A manufacturer's announcement to stop all work using faulty equipment

   c. A site contractor's decision not to use particular materials

   d. A local authority banning the use of a particular type of brick

3. Which of the following is considered a major injury?

   a. Bruising on the knee

   b. Cut

   c. Concussion

   d. Exposure to fumes

4. If there is an accident on a site who is likely to be the first to respond?

   a. First aider

   b. Police

   c. Paramedics

   d. HSE

5. Which of the following is a summary of risk assessments and is used for high risk activities?

   a. Site notice board

   b. Hazard book

   c. Monitoring statement

   d. Method statement

6. Some substances are combustible. Which of the following are examples of combustible materials?

   a. Adhesives

   b. Paints

   c. Cleaning agents

   d. All of these

7. What is dermatitis?

   a. Inflammation of the skin

   b. Inflammation of the ear

   c. Inflammation of the eye

   d. Inflammation of the nose

8. Screens, netting and guards on a site are all examples of which of the following?

   a. PPE

   b. Signs

   c. Perimeter safety

   d. Electrical equipment

9. Which of the following are also known as scissor lifts or cherry pickers?

   a. Bench saws

   b. Hand-held power tools

   c. Cement additives

   d. Mobile elevated work platforms

10. In older properties the neutral electricity wire is which colour?

    a. Black

    b. Red

    c. Blue

    d. Brown

Chapter 2

Unit CSA–L2Core04

UNDERSTAND INFORMATION, QUANTITIES AND COMMUNICATION WITH OTHERS

LEARNING OUTCOMES

LO1: Know how to interpret and produce information relating to construction

LO2: Understand how to estimate quantities of resources

LO3: Understand how to communicate workplace requirements efficiently

INTRODUCTION

The aim of this chapter is to:

* help you interpret and produce information relating to construction

* show you how to estimate quantities of resources

* enable you to communicate workplace requirements effectively to all levels of the construction team.

INTERPRETING AND PRODUCING INFORMATION

Even quite simple construction projects will require documents. These provide you with the necessary information that you will need to do the job. The documents are produced by a range of different people and each document has a different purpose. Together they give you the full picture of the job, from the basic outline through to the technical specifications.

Types of supporting information

Supporting information can be found in a variety of different types of documents. These include:

* drawings and plans

* programmes of work

* procedures

* specifications

* policies

* schedules

* manufacturers' technical information

* organisational documentation

* training and development records

* risk and method statements

* Construction (Design and Management) (CDM) Regulations

* Building Regulations.

Drawings and plans

Drawings are an important part of construction work. You will need to understand how drawings provide you with the information required to carry out the work. The drawings show what the building will look like and how it will be constructed. This means that there are several different drawings of the building from different viewpoints. In practice, most of the drawings are shown on the same sheet.

Block plans

Block plans show the construction site and the surrounding area. Normally block plans are at a ratio of 1:2500 and 1:1250. This means that 1 millimetre on a block plan is equal to 2,500 mm (2.5 m) or 1,250 mm (1.25 m) on the ground.

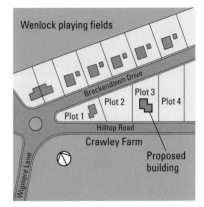

Figure 2.1 Block plan

Site plan

Location drawings are sometimes known as site plans. The site plan drawing shows what is basically planned for the site. It is an important drawing because it has been created in order to get approval for the project from planning committees or funding sources. In most cases the site plan is actually an architectural plan, showing the basic arrangement of buildings and any landscaping.

The site plan will usually show:

* directional orientation (i.e. the north point)

* location and size of the building or buildings

* existing structures

* clear measurements

* colours and materials to be used.

General location

Location drawings show the site or building in relation to its surroundings. It will therefore show details such as boundaries, other buildings and roads. It will also contain other vital information, including:

* access

* drainage

* sewers

* the north point.

The drawing will have a title and will show the scale. A job or project number will help to identify it easily, and it will also have an address, the date when the drawing was done and the name of the client. A version number will also be on the drawing, with an amendment date if there have been any changes. It is important to make sure you have the latest drawing.

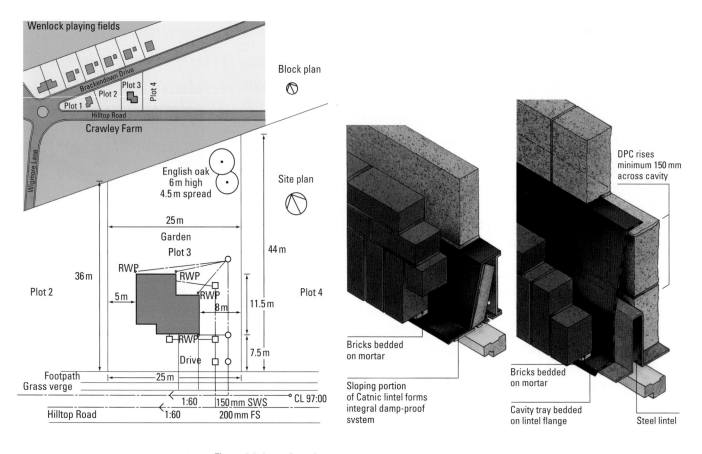

Figure 2.2 Location plan

Figure 2.3 Assembly drawing

Normally location drawings are either 1:500 or 1:200 (that is, 1 mm of the drawing represents 500 mm or 200 mm on the ground).

Assembly

These are detailed drawings that illustrate the different elements and components of the construction. They are likely to be 1:20, 1:10 or 1:5 (1 mm of the drawing represents 20 mm, 10 mm or 5 mm on the ground). This larger scale allows more detail to be shown, to ensure accurate construction.

Sectional

These drawings aim to provide:

* vertical dimensions

* constructional details

* horizontal sections.

They can be used to show the height of ground levels, damp-proof courses, foundations and other aspects of the construction.

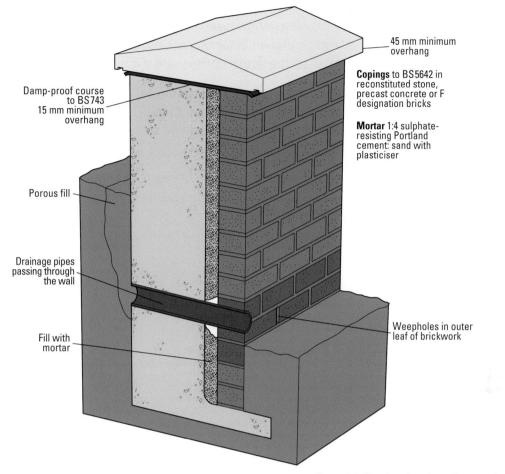

Damp-proof course to BS743 15 mm minimum overhang

Porous fill

Drainage pipes passing through the wall

Fill with mortar

45 mm minimum overhang

Copings to BS5642 in reconstituted stone, precast concrete or F designation bricks

Mortar 1:4 sulphate-resisting Portland cement: sand with plasticiser

Weepholes in outer leaf of brickwork

Figure 2.4 Section drawing of an earth retaining wall

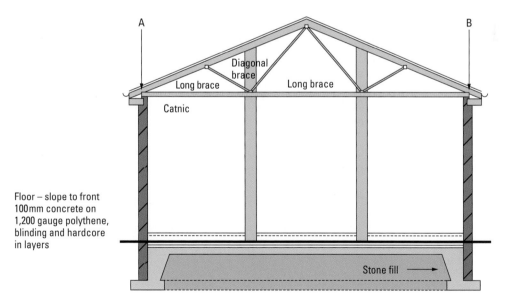

A   B

Diagonal brace

Long brace   Long brace

Catnic

Floor – slope to front 100mm concrete on 1,200 gauge polythene, blinding and hardcore in layers

Stone fill →

Figure 2.5 Section drawing of a garage

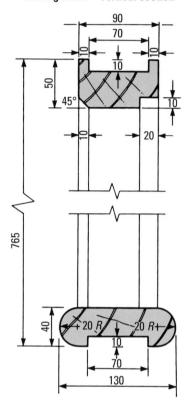

Serving hatch   Vertical section

Figure 2.6 Detail drawing

Details

These drawings show how a component needs to be manufactured. They can be shown in various scales, but mainly 1:10, 1:5 and 1:1 (the same size as the actual component if it is small).

Orthographic projection (first angle)

First angle projection is a view that represents the side of the object as if you were standing away from it, as can be seen in Fig 2.7.

Isometric projection

Isometric projection is a way of representing three-dimensional objects in two dimensions, as can also be seen in Fig 2.7. All horizontal lines are drawn at 30°.

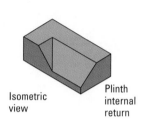

Isometric view

Plinth internal return

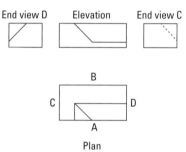

End view D   Elevation   End view C

B

C   D

A

Plan

Figure 2.7 First angle projection

Programmes of work

Programmes of work show the actual sequence of any work activities on a construction project. Part of the work programme plan is to show target times. They are usually shown in the form of a bar or Gantt chart (a special kind of bar chart), as can be seen in Fig 2.8.

In this figure:

* on the left hand side all of the tasks are listed – note this is in logical order

* on the right the blocks show the target start and end date for each of the individual tasks

* the timescale can be days, weeks or months.

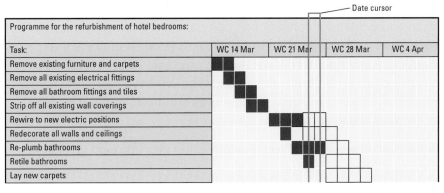

Date cursor

Programme for the refurbishment of hotel bedrooms:				
Task:	WC 14 Mar	WC 21 Mar	WC 28 Mar	WC 4 Apr
Remove existing furniture and carpets				
Remove all existing electrical fittings				
Remove all bathroom fittings and tiles				
Strip off all existing wall coverings				
Rewire to new electric positions				
Redecorate all walls and ceilings				
Re-plumb bathrooms				
Retile bathrooms				
Lay new carpets				

Figure 2.8 Single line contract plan Gantt chart

Far more complex forms of work programmes can also be created. The Gantt chart shown below (Fig 2.9) shows the construction of a house.

This is a more complex example of a bar chart:

* There are two lines – they show the target dates and actual dates. The actual dates are shaded, showing when the work actually began and how long it actually took.

* If this bar chart is kept up to date an accurate picture of progress and estimated completion time can be seen.

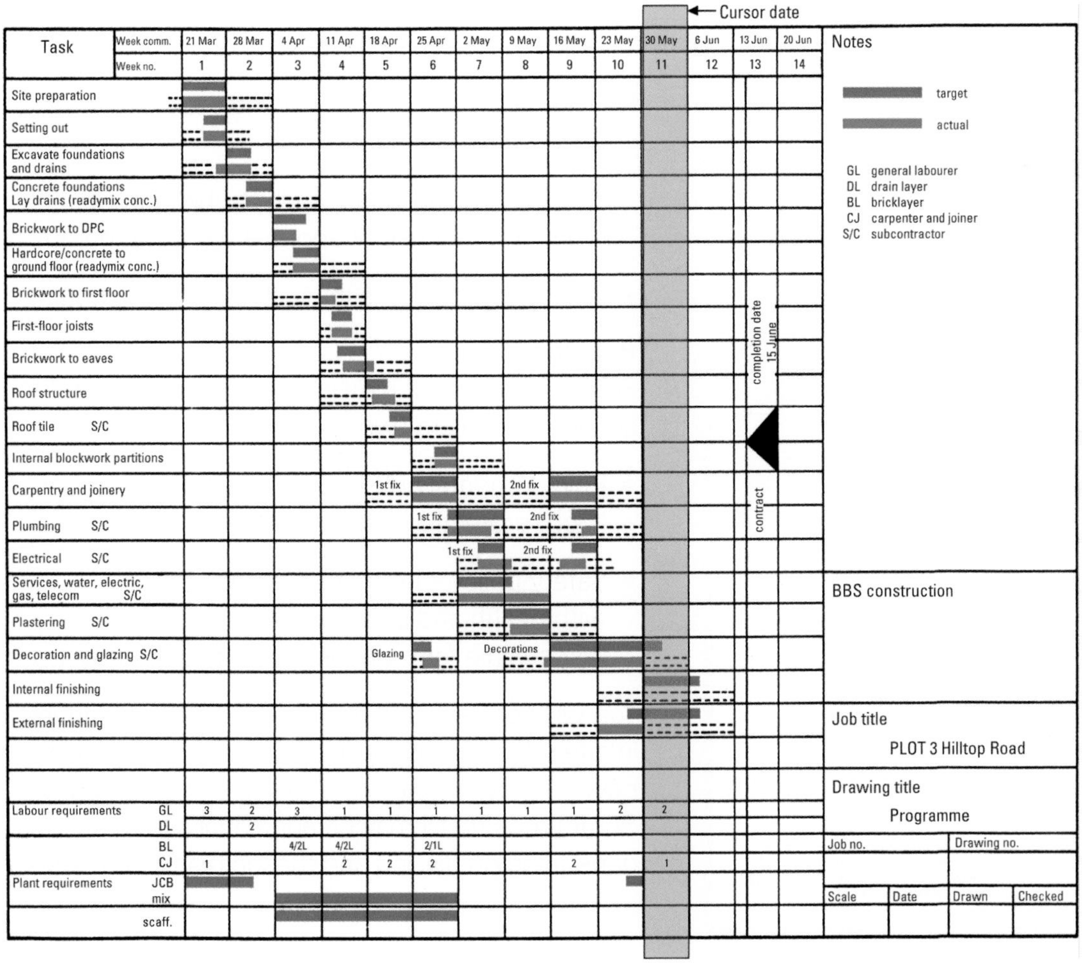

← Cursor date

Figure 2.9 Gantt chart for the construction of a house

Procedures

When you work for a construction company it will have a series of procedures which you will have to follow. A good example is the emergency procedure. This will explain precisely what is required in the case of an emergency on site and who will have responsibility for carrying out particular duties. Procedures are there to show you the right way of doing something.

Another good example of a procedure is the procurement or buying procedure. This will outline:

* who is authorised to buy what, and how much individuals are allowed to spend

* any forms or documents that have to be completed when buying.

Specifications

In addition to drawings it is usually necessary to have documents known as specifications. These provide much more information, as can be seen in Fig 2.10.

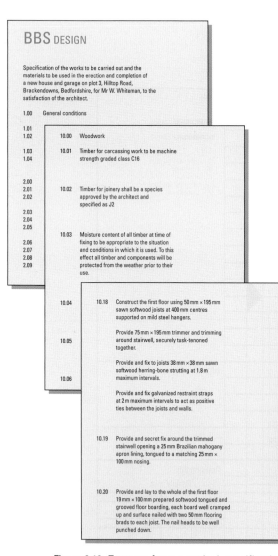

The specifications give you a precise description. They will include:

* the address and description of the site

* on-site services (e.g. water and electricity)

* materials description, outlining the size, finish, quality and tolerances

* specific requirements, such as the individual who will authorise or approve work carried out

* any restrictions on site, such as working hours.

Policies

Policies are sets of principles or a programme of actions. These are two good examples:

* Environmental policy – how the business goes about protecting the environment.

* Safety policy – how the business deals with health and safety matters and who is responsible for monitoring and maintaining it.

You will normally find both policies and procedures in site rules. These are usually explained to each new employee when they first join the company. Sometimes there may be additional site rules, depending on the job and the location of the work.

Figure 2.10 Extracts from a typical specification

Schedules

Schedules are cross-referenced to drawings that have been prepared by an architect. They will show specific design information. Usually they are prepared for jobs that will be carried out regularly on site, such as:

* working on windows, doors, floors, walls or ceilings

* working on drainage, lintels or sanitary ware.

A schedule can be seen in Fig 2.11.

The schedule is very useful for a number of purposes, such as:

* working out the quantities of materials needed

* ordering materials and components and then checking them against deliveries

* locating where specific materials will be used.

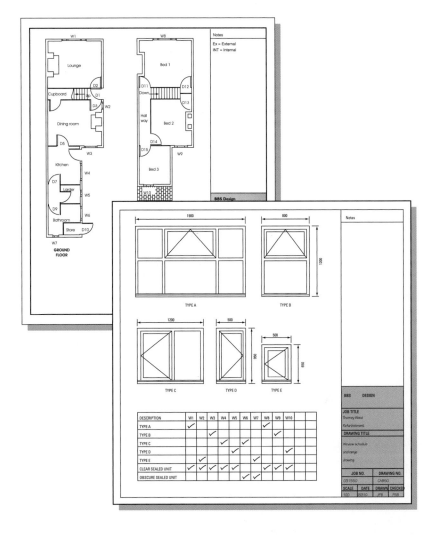

Figure 2.11 Typical windows schedule, range drawing and floor plans

Manufacturers' technical information

Almost everything that is bought to be used on site will come with a variety of types of information. The basic technical information provided will show what the equipment or material is intended to be used for, how it should be stored and any particular requirements it may have, such as for handling or maintenance.

Technical information from the manufacturer can come from a variety of different sources. These may include:

* printed or downloadable data sheets

* printed or downloadable user instructions

* manufacturers' catalogues or brochures

* manufacturers' websites.

Organisational documentation

The potential list of organisational documentation and paperwork is extensive. These are outlined in Table 2.1. Examples can be seen in Figs 2.12 to 2.16.

Document	Purpose
Timesheet	Record of hours that you have worked and the jobs that you have carried out. This is used to help work out your wages and the total cost of the job.
Day worksheet	This details work that has been carried out without providing an estimate beforehand. It usually includes repairs or extra work and alterations.
Variation order	Provided by the architect and given to the builder, showing any alterations, additions or omissions to the original job.
Confirmation notice	Provided by the architect to confirm any verbal instructions.
Daily report or site diary	This covers things that might affect the project like detailed weather conditions, late deliveries or site visitors.
Orders and requisitions	These are order forms, requesting the delivery of materials.
Delivery notes	These are provided by the supplier of materials as a list of all materials being delivered. These need to be checked against materials actually delivered. The buyer will sign the delivery note when they are happy with the delivery.
Delivery records	These are lists of all materials that have been delivered on site.
Memorandum	These are used for internal communications and are usually brief.
Letters	These are used for external communications, usually to customers or suppliers.
Fax	Even though email is commonly used, the industry still uses faxes, as they provide an exact copy of an original document.

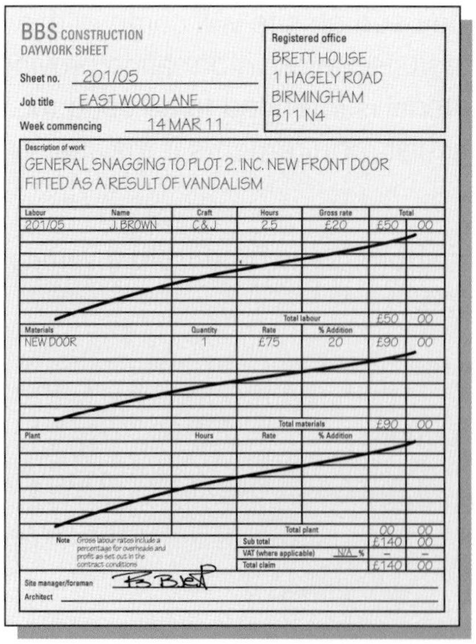

Table 2.1 Types of organisational documentation

BBS CONSTRUCTION
WEEKLY TIME SHEET

Registered office: BRETT HOUSE, 1 HAGELY ROAD, BIRMINGHAM, B11 N4

Name: JAMES BROWN
Craft: CARPENTER
Week commencing: 14 MAR 11

	Job title	Description of work	Time: start/finish	total
MON	EAST WOOD LANE	SECOND FIXING PLOT 10	8.00 TO 5.00	8.5
TUE	EAST WOOD LANE	SECOND FIXING PLOT 10	8.00 TO 5.00	8.5
WED	EAST WOOD LANE	SECOND FIXING PLOT 11	8.00 TO 5.00	8.5
THUR	EAST WOOD LANE	HANGING EXTERNAL DOORS PLOT 17	8.00 TO 5.00	8.5
FRI	EAST WOOD LANE	FIXING DOOR FURNITURE AND GENERAL SNAGGING FOR HANDOVER PLOT 2	8.00 TO 3.00	6.5
SAT				
SUN				

Details of expenses (attach receipts): N/A
Authorized by: [signature] Position: PROJECT MANAGER
For office use only
Standard hours ___ at ___ = ___
Overtime hours ___ at ___ = ___
Overtime hours ___ at ___ = ___
Overtime hours ___ at ___ = ___
*A half-hour lunch break must be deducted from all daily totals. TOTAL = ___

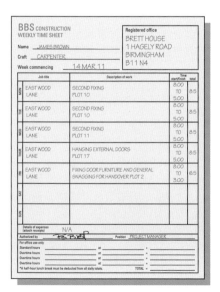

Figure 2.12 Timesheet

BBS CONSTRUCTION
DAYWORK SHEET

Registered office: BRETT HOUSE, 1 HAGELY ROAD, BIRMINGHAM, B11 N4

Sheet no.: 201/05
Job title: EAST WOOD LANE
Week commencing: 14 MAR 11

Description of work: GENERAL SNAGGING TO PLOT 2. INC. NEW FRONT DOOR FITTED AS A RESULT OF VANDALISM

Labour	Name	Craft	Hours	Gross rate	Total
201/05	J. BROWN	C&J	2.5	£20	£50 00
			Total labour		£50 00

Materials	Quantity	Rate	% Addition	Total
NEW DOOR	1	£75	20	£90 00
		Total materials		£90 00

Plant	Hours	Rate	% Addition	Total
		Total plant		00 00

Note: Gross labour rates include a percentage for overheads and profit as set out in the contract conditions

Sub total: £140 00
VAT (where applicable) N/A %: – –
Total claim: £140 00

Site manager/foreman: [signature]
Architect: ___

Figure 2.13 Day worksheet

Variation Order

Project: ALFA HOUSE REFURBISHMENTS
Location: NOTTINGHAM
Client: CREASTA HOTELS GROUP
Project No: 14/035

CAB ARCHITECTS
CAB ARCHITECTS LTD, BRETT HOUSE, BEASTON, NOTTINGHAM, NG10 2XF
TEL: 0115 9000010
FAX: 0115 9000011
EMAIL: CAB@ARCH.COM

From: JAMES BLUNT To: DAVID SMURTHWAITE
Position: PROJECT DESIGNER Position: M.D. HOTELS

Additions:
PROVIDE ADJUSTABLE SHELVING TO BACK BAR FITTING AS DETAILED ON DRAWING NO. 14/0356 REV B.

Omissions:
DELETE FIXED SHELVING TO BACK BAR FITTING DETAILED ON DRAWING NO. 14/0356 REV A.

Variation Order No.: CHG/VO/004
Approved by: [signature]
Date: 15 MARCH 2011

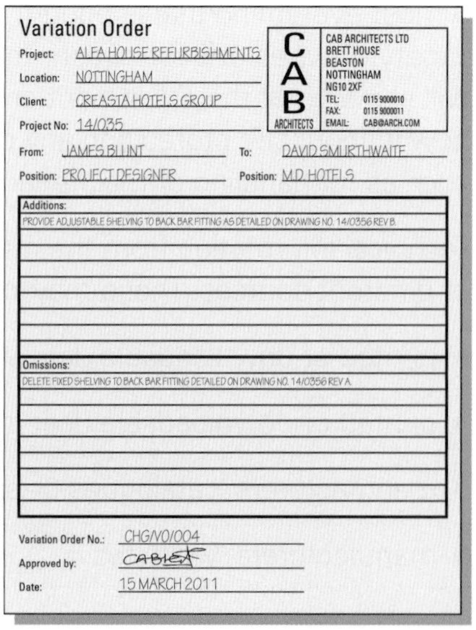

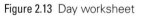

Figure 2.14 Variation order

Training and development records

Training and development is an important part of any job, as it ensures that employees have all the skills and knowledge that they need to do their work. Most medium to large employers will have training policies that set out how they intend to do this.

Employers will have a range of different documents to keep records and to make sure that they are on track. These documents will record all the training that an employee has undertaken.

Training can take place in a number of different ways and different places. It can include:

* induction

* toolbox talks

* in-house training

* specialist training

* training or education leading to formal qualifications.

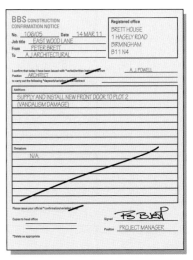

Figure 2.15 Confirmation notice

Figure 2.16 Daily report or site diary

Checking information for conformity

The information to be checked can include drawings, programmes of work, schedules, policies, procedures, specifications and so on. The term 'conformity' in this sense means:

* making sure that any part of the assembly or component is suitable for the job

* making sure that the standard of work meets the necessary performance requirements.

This may mean that there could be an industry or trade standard that will need to be followed. The actual job or client may also require specific standards.

Interpreting construction specifications

It would be difficult to put in all of the details in full, so symbols, hatchings and abbreviations are used to simplify the drawings. All of these symbols or hatchings are drawn to follow BS1192. The symbols cover various types of brickwork and blockwork, as well as concrete, hard core and insulation, as can be seen in Fig 2.17.

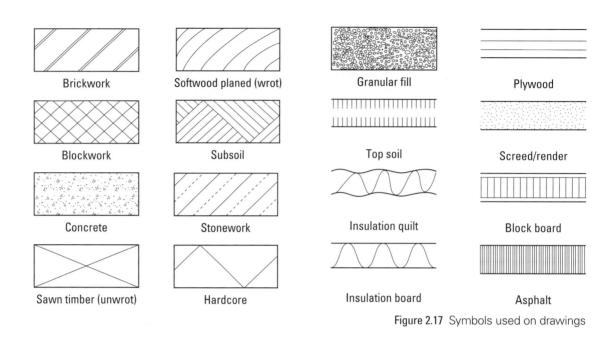

Figure 2.17 Symbols used on drawings

Common abbreviations

As we have already seen, covering the drawing with full detail would make it hard to read, so abbreviations are used. Table 2.2 outlines some examples of abbreviations that you will need to become familiar with.

Abbreviation	Meaning
bwk	Refers to all types of brickwork
conc	Refers to areas that will be concreted
dpc	Refers to all types of damp-proof course
fdn	Refers to foundations that are required
insul	Refers to location and type of insulation
rwg	Refers to location of rainwater gulleys
svp	Refers to location and type of soil and vent pipe

Table 2.2 Abbreviations used in construction drawings

Drawing equipment and its uses

Some basic equipment is necessary in order to produce drawings. These items are outlined in Table 2.3.

Equipment	Explanation and use
Scale rule	This is an essential piece of equipment. It needs to have 1:5/1:50, 1:10/1:100, 1:20/1:200 and 1:250/1:2500.
Set square	You will need to have a pair of these, or an adjustable square. If it is adjustable then you need to be able to create angles of up to 90°. The set square on the shortest side should be at least 150 mm. You will need the ability to create 30, 45, 60 and 90° angles.
Protractor	A protractor is essential to be able to measure angles up to and including 180°.
Compass	Compasses are used to create circles or arcs. It is also advisable to have a divider so that you can easily transfer measurements and dividing lines.
Pencils	For drawings you will need a 2H, 3H or 4H pencil. For sketching and darkening outlines you will need an HB pencil. You will need to keep these sharp.

Table 2.3 Drawing equipment required

In addition to this you will also need at least an A2 size drawing board that has a parallel rule (these may be provided by your college). It is also useful to have an eraser.

Scales used to produce construction drawings

When the plans for individual buildings or construction sites are drawn up they have to be scaled down so that they will fit on a manageable size of paper. It is important to remember that drawings are not sketches and that they are drawn to scale. This means that they are:

* exact and accurate

* in proportion to the real construction.

You can work out the dimensions by using the scale rule when measuring the drawings. There are several common scales used and the measurement is usually metric:

* 1:2500 – the drawing is 2,500 times smaller than the real object

* 1:100 – the drawing is 100 times smaller than the real object

* 1:50 – the drawing is 50 times smaller than the real object

* 1:20 – the drawing is 20 times smaller than the real object

* 1:10 – the drawing is 10 times smaller than the real object

* 1:5 – the drawing is 5 times smaller than the real object

* 1:2 – the drawing is 2 times smaller than the real object.

ESTIMATING QUANTITIES OF RESOURCES

Working out the quantity and cost of resources that are needed to do a particular job is, perhaps, one of the most difficult tasks. In most cases you or the company you work for will be asked to provide a price for the work.

It is generally accepted that there are three ways of doing this:

* estimate – an approximate calculation based on available information

* quotation – which is a fixed price

* tender – tendering is a process of allowing various parties to price for the same work. The process can be open or closed. This usually means that the result is fair.

As we will see a little later in this section, these three ways of costing are very different and each of them has its own problems.

Methods used to estimate quantities

Obviously past experience will help you to quickly estimate the amount of materials that will be needed on particular construction projects. This is also true of working out the best place to buy materials and how much the labour costs will be to get the job finished.

Many businesses will use the *Hutchins UK Building Costs Blackbook*, which provides a construction cost guide. It breaks down all types of work and shows an average cost for each of them.

Computerised estimating packages are available, which will give a comprehensive detailed estimate that looks very professional. This will also help to estimate quantities and timescales.

The alternative is of course to carry out a numerical calculation. It is therefore important to have the right resources upon which to base these calculations. These could be working drawings, schedules or other documents.

Usually all this involves making additions, subtractions, multiplications and divisions. In order to work out the amount of materials you will need for a construction project you will need to know some basic information:

* What does the job entail? How complex is it, and how much labour is required?

* What materials will be used?

* What are the costs of the materials?

Measurement

The standard unit for measurement is the metre (m). There are 100 centimetres (cm) and 1,000 millimetres (mm) in a metre. It is important to remember that drawings and plans have different scales, so these need to be converted to work out the quantities of materials required.

The most basic thing to work out is length (see Fig 2.18), from which you can calculate perimeter, area and then volume, capacity, mass and weight, as can be seen in Table 2.4.

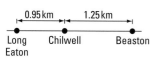

Long lengths in kilometres (km)

Intermediate lengths in metres (m)

75 mm

Small lengths in millimetres (mm)

Figure 2.18 Length in metres and millimetres

Measurement	Explanation
Length	This is the distance from one end to the other. This could be measured in metres or millimetres, depending on the job.
Perimeter	This helps you work out the distance around a shape, such as the size of a room or a garden. It will help you estimate the length of a wall, for example. You just need to measure each side and then add them together (see Fig 2.20).
Area	You can work out the area of a room, for example, by measuring the length and the width of the room. Then you multiply the width by the length to give the number of square metres (m²) (see Fig 2.20).
Volume and capacity	Volume shows how much space is taken up by an object, such as a room. Again this is simply worked out by multiplying the width of the room by its length and then by its height. This gives you the number of cubic metres (m³). Capacity works in exactly the same way but instead of showing the figure as cubic metres you show it as litres. This is ideal if you are trying to work out the capacity of a water tank or a garden pond (see Fig 2.19).
Mass or weight	Mass is measured usually in kilograms or in grammes. Mass is the actual weight of a particular object, such as a brick.

Table 2.4 Working out measurements

Shape	Area equals	Perimeter equals
Trapezium	$\dfrac{(A + B)H}{2}$ (or A plus B multiplied by H and then divided by 2)	A + B + C + D
Triangle	$\dfrac{BH}{2}$ (or B multiplied by H and then divided by 2)	A + B + C
Circle	πr^2 (or r multiplied by itself and then multiplied by pi (3.142))	πd or $2\pi r$

Table 2.5 Calculating complex areas

Volume

Sometimes it is necessary to work out the volume of an object, such as a cylinder or the amount of concrete needed. All that needs to be done is to work out the base area and then multiply that by the height.

For a concrete volume, if a 1.2 m square needs 3 m of height then the calculation is:

$$1.2 \times 1.2 \times 3 = 4.32\,\text{m}^3$$

To work out the volume of a cylinder you need to know the base area × the height. The formula is:

$$\pi r^2 \times H$$

So if a cylinder has a radius (r) of 0.8 and a height of 3.5 m then the calculation is:

$$3.142 \times 0.8 \times 0.8 \times 3.5 = 7.038\,\text{m}^3$$

Pythagoras

Pythagoras' theorem is used to work out the length of the sides of right-angled triangles. It states that:

In all right-angled triangles the square of the longest side is equal to the sum of the squares of the other two sides (that is, the length of a side multiplied by itself).

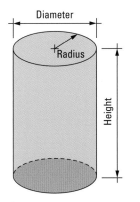

Figure 2.24 Cylinder

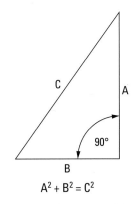

$$A^2 + B^2 = C^2$$

Figure 2.25 Pythagoras' theorem

Measuring materials

Using simple measurements and formulae can help you work out the amount of materials you will need. This is all summarised in Table 2.6.

Material	Measurement
Timber	To work out the linear run of a cubic metre of timber of a given cross sectional area, divide a square metre by the cross sectional area of one piece.
Flooring	To work out the amount of flooring for a particular area in metres^2 multiply the width of the floor by the length of the floor.
Stud walling, rafters and joists	Measure the distance that the stud partition will cover then divide that distance by a specified spacing and add 1. This will give you the number of spaces between each stud.
Fascias, barges and soffits	Measure the length and then add 10% for waste; however, this will depend on the nearest standard metric size of timber available.
Skirting, dado, picture rails and coving	You need to work out the perimeter of the room and then subtract any doorways or other openings. Again, add 10% for waste.
Bricks and mortar	Half brick walls use 60 bricks per metre squared and one brick walls use double that amount. You should add 5 per cent to take into account any cutting or damage. For mortar assume that you will need 1 kg for each brick.

Table 2.6 Working out materials required

How to cost materials

Once you have found out the quantity of materials necessary, you need to find out the price of those materials. You then do the costing by simply multiplying those prices by the amount of materials actually needed.

CASE STUDY

South Tyneside Homes

South Tyneside Council's Housing Company

It's important to get it right

Glen Campbell is a team leader at South Tyneside Homes.

'Your English and maths skills really are important. As an apprentice, you have to be able to communicate properly – to get information and materials back and forth between tradespeople and yourself, to be able to sit and put a little drawing down, to label things up, and to take information off drawings – especially on the capital works jobs. You're reading and writing stuff down all the time… even your timesheets because they have to be accurate.

When it comes to your maths skills, you're using measurement all the time. If you get measurements wrong, you're not making the money. For example, if you're using the wrong size timber for a roof – the drawing says you've got to use 200 × 50 mm joists and then you go and use ones that are 150 mm – it's either going to cost you more to go back and get it right, or it's not going to be able to take that stress load once the roof goes on. In the end it could even collapse.'

Materials and purchasing systems

Many builders and companies will have preferred suppliers of materials. Many of them will already have negotiated discounts based on their likely spending with that supplier over the course of a year. The supplier will then be organised to supply them at an agreed price.

In other cases, builders may shop around to find the best price for the materials that match the specification. The lowest price may not necessarily be the best one to go for. All materials need to be of a sufficient quality. The other key consideration is whether the materials are immediately available for delivery.

It is vital that suppliers are reliable and that they have sufficient materials in stock. Delays in deliveries can cause major setbacks on site. It is not always possible to warn suppliers that materials will be needed, but a well-run site should be able to anticipate the materials that are needed and put in the orders within good time.

Large quantities may be delivered direct from the manufacturer straight to site. This is preferable when dealing with items where colours must be consistent.

Comparing estimated labour rates

The cost of labour for particular jobs is based on the hourly charge-out rate for that individual or group of individuals multiplied by the time it would take to complete the job.

Labour rates can depend on:

* the expertise of the construction worker
* the size of the business they work for
* the part of the country in which the work is being carried out
* the complexity of the work.

According to the International Construction Costs Survey 2012, the following were average costs per hour:

* Group 1 tradespeople – plumbers, electricians etc. – £30
* Group 2 tradespeople – carpenters, bricklayers etc. – £30
* Group 3 tradespeople – tillers, carpet layers and plasterers – £30
* General labourers – £18
* Site supervisors – £46.

Quotes, estimated prices and tenders

As we have already seen, estimates, quotes and tenders are very different. It is useful to look at these in slightly more detail, as can be seen in Table 2.7 below.

Type of costing	Explanation
Estimate	This needs to be a realistic proposal of how much a job will cost. An estimate is not binding and the client needs to understand that the final cost might be more.
Quote	This is a fixed price based on a fixed specification. The final price may be different if the fixed specification changes, for example if the customer asks for additional work then the price will be higher.
Tender	This is a competitive process. The customer advertises the fact that they want a job done and invites tenders. The customer will specify the specifications and schedules and may even provide the drawings. The companies tendering then prepare their own documents and submit their price based on the information the customer has given them. All tenders are submitted to the customer by a particular date and are either open or closed. The customer then opens all tenders on a given date and awards the contract to the company of their choice. This process is particularly common among public sector customers, such as local authorities.

Table 2.7 Estimates, quotes and tenders

Implications of inaccurate estimates

Larger companies will have an estimating team. Smaller businesses will have someone who has the job of being an estimator. Whenever they are pricing a job, whether it is a quote, an estimate or a tender, they will have to work out the costs of all materials, labour and other costs. They will also have to include a **mark-up**.

It is vital that all estimating is accurate. Everything needs to be measured and checked. All calculations need to be double-checked.

It can be disastrous if these figures are wrong because:

* if the figure is too high then the client is likely to reject the estimate and look elsewhere as some competitors could be cheaper

* if the figure is too low then the job may not provide the business with sufficient profit and it will be a struggle to make any money out of the job.

KEY TERMS

Mark-up

– a builder or building business, just like any other business, needs to make a profit. Mark-up is the difference between the total cost of the job and the price that the customer is asked to pay for the work.

DID YOU KNOW?

Many businesses fail as a result of not working out their costs properly. They may have plenty of work but they are making very little money.

COMMUNICATING WORKPLACE REQUIREMENTS

Communication can be split into two different types:

* Verbal communication – including face-to-face conversations, discussions in meetings or performance reviews and talking on the telephone.

* Written communication – including all forms of documents, from letters and emails to drawings and work schedules.

Each of these forms of communication needs to be clear, accurate and designed in such a way as to make sure that whoever has to use it or refer to it understands it.

Figure 2.26 It's important to communicate effectively, whether it's verbal or written

Key personnel in the communication cycle

Each construction job will require the services of a team of professionals. They have to be able to work and communicate effectively with one another. Each team has different roles and responsibilities. They can be broken down into three particular groups:

* on site   * off site   * visitors.

These are described in Tables 2.8, 2.9 and 2.10.

Role	Responsibilities
Apprentices	They can work for any of the main building services trades under supervision. They only carry out work that has been specifically assigned to them by a trainer, a skilled operative or a supervisor.
Skilled or trade operative	A specialist in a particular trade, such as bricklaying or carpentry. They will be qualified in that trade, or working towards their qualification
Unskilled operatives	Also known as labourers, these are entry level operatives without any formal training. They may be experienced on sites and will take instructions from the supervisor or site manager.
Building services engineers	They are involved in the design, installation and maintenance of heating, water, electrics, lighting, gas and communications. They work either for the main contractor or the architect and give instruction to building services operatives.
Building services operatives	They include all the main trades involved in installation, maintenance and servicing. They take instruction from the building services engineers and work with other individuals, such as the supervisor and charge-hand.
Charge-hand	This person supervises a specific trade, such as carpenters and bricklayers.
Trade foreperson	This person supervises the day-to-day running of the site, and organises the charge-hand and any other operatives.
Site manager	This person runs the construction site, makes plans to avoid problems and meet deadlines, and ensures all processes are carried out safely. They communicate directly with the client.
Supervisor	The supervisor works directly for the site manager on larger projects and carries out some of the site manager's duties on their behalf.
Health and safety officer	This person is responsible for managing the safety and welfare of the construction site. They will carry out inspections, provide training and correct hazards.

Table 2.8 On-site construction team

Role	Responsibilities
Client	The client, such as a local authority, commissions the job. They define the scope of the work and agree on the timescale and schedule of payments.
Customer	For domestic dwellings, the customer may be the same as the client, but for larger projects a customer may be the end user of the building, such as a tenant renting local authority housing or a business renting an office. These individuals are most affected by any work on site. They should be considered and informed with a view to them suffering as little disruption as possible.
Architect	They are involved in designing new buildings, extensions and alterations. They work closely with clients and customers to ensure the designs match their needs. They also work closely with other construction professionals, such as surveyors and engineers.
Consultant	Consultants such as civil engineers work with clients to plan, manage, design or supervise construction projects. There are many different types of consultant, all with particular specialisms.
Main contractor	This is the main business or organisation employed to head up the construction work. The contractor organises the on-site building team and pulls together all necessary expertise. They manage the whole project, taking full responsibility for its progress and costs.
Clerk of works	This person is employed by the architect on behalf of a client. They oversee the construction work and ensure that it represents the interests of the client and follows agreed specifications and designs.
Quantity surveyor	Quantity surveyors are concerned with building costs. They balance maintaining standards and quality against minimising the costs of any project. They need to make choices in line with Building Regulations. They may work either for the client or for the contractor.
Estimator	Estimators calculate detailed cost breakdowns of work based on specifications provided by the architect and main contractor. They work out the quantity and costs of all building materials, plant required and labour costs.
Sub-contractor	They carry out work on behalf of the main contractor and are usually specialist tradespeople or professionals, such as electricians. Essentially, they provide a service and are contracted to complete their part of the project.
Supplier/wholesaler contracts manager	They work for materials suppliers or stockists, providing materials that match required specifications. They agree prices and delivery dates.

Table 2.9 Off-site construction team

Site visitor	Role and responsibility
Training officers and assessors	These people work for approved training providers. They visit the site to observe and talk to apprentices and their mentors or supervisors. They assess apprentices' competence and help them to put together the paperwork needed to show evidence of their skills.
Building control inspector	This person works for the local authority to ensure that the construction work conforms to regulations, particularly the Building Regulations. They check plans, carry out inspections, issue completion certificates, work with architects and engineers and provide technical knowledge on site.
Water inspector	This person carries out checks of plumbing and drainage systems on construction sites.
Health and Safety Executive (HSE) inspector	An HSE inspector can enter any workplace without giving notice. They will look at the workplace, the activities and the management of health and safety to ensure that the site complies with health and safety laws. They can take action if they find there is a risk to health and safety on site.
Electrical services inspector	Inspectors are approved by the National Inspection Council for Electrical Installation Contracting. They check all electrical installation has been carried out in accordance with legislation, particularly Part P of the Building Regulations.

Table 2.10 Construction visitors

Effects of poor communication

Effective communication is essential in all types of work. It needs to be clear and to the point, as well as accurate. Above all it needs to be a two-way process. This means that any communication that you have with anyone must be understood by them. It means thinking before communicating. Never assume that someone understands you unless they have confirmed that they do.

In construction work you have to keep to schedule and work on time, and it is important to follow precise instructions and specifications. Failing to communicate will always cause confusion, extra cost and delays; it can lead to problems with health and safety and accidents. Such problems are unacceptable and very easy to avoid. Negative communication or poor communication can damage the confidence that others have in you to do your job.

Good communication means efficiency and achievement.

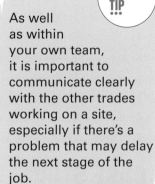

REED TIP

As well as within your own team, it is important to communicate clearly with the other trades working on a site, especially if there's a problem that may delay the next stage of the job.

Communication techniques and teamwork

It is important to have a good working relationship with colleagues at work. An important part of this is to communicate in a clear way with them. This helps everyone understand what is going on and what decisions have been made. It also means being clear. Most communication with colleagues will be verbal (spoken). Good communication means:

* cutting out mistakes and stoppages (saving money)

* avoiding delays

* making sure that the job is done right the first time and every time.

Figure 2.27 A water inspection

Equality and diversity in communication

Equality and diversity is not simply about treating everyone in the same way. It is actually recognising that people are different and have different needs. Each of us is unique. This could mean that you are working with people of a different culture, a different age (younger or older), or who follow different religions. It might refer to marital status or gender, sexual orientation or your first language.

In all your actions and your communications you should:

* recognise and respect other people's backgrounds

* recognise that everyone has rights and responsibilities

* not harass or be offensive and use language or behaviour that discriminates.

You should also remember that not everyone's first language will be English so they may not understand everything or be able to communicate clearly with you. You might also find that some colleagues may have hearing impairments (or may not hear what you're saying because they are in a noisy environment). In cases like these, use simple language and check that both you and the person you are communicating with have understood the message.

CASE STUDY

South
Tyneside *Homes*

South Tyneside Council's
Housing Company

Using writing and maths in the real world

Gary Kirsop, Head of Property Services, says:

'People seem to think that trades are all about your hands, but it's more than that. You're measuring complicated things – all the trades need to have about the same technical level for planning, calculation and writing reports. You need that level to get through your exams for the future too. When you have one day a week in college, but four days a week working with customers in the real world, without communications skills, it would all fall apart. You have to understand that people come from different backgrounds and that they have their own communication modes. Having good GCSEs will really help you get by in the trade.'

Advantages and disadvantages of different methods of communication

As you progress in your career in construction, you may come across a number of different documents that are used either in the workplace or are provided to customers or clients. All of these documents have a specific purpose. Their exact design may vary from business to business, but the information contained on them will usually be similar.

Documents in the workplace

This group of documents tend to be used only within the workplace. Their general purpose is to collect information or to pass on information from one part of the business to another.

Document type	Purpose
Job specifications	These are detailed sets of requirements that cover the construction, features, materials, finishes and performance specifications required for each major aspect of a project. They may, for example, require a particular level of energy efficiency.
Plans or drawings	These are prepared by architects. They are drawn to scale and provide a standard detailed drawing. They will be used as blueprints (instructions) by building services engineers and operatives while they are working on the site.
Work programmes	These are detailed breakdowns of the order in which work needs to be completed, along with an estimate as to how long each stage is likely to take. For example, a certain amount of time will be allocated for site preparation and then piling and the construction of the substructure of the building. The work programme will indicate when particular skills will be needed and for approximately how long.
Purchase orders	These are documents issued by the buyer to a supplier. They detail the type of materials, quantity and the agreed price so form a record of what has been agreed. The order for materials will have been discussed with the supplier before the purchase order is completed. Many purchase orders are now transmitted electronically, although paper records may be necessary for future reference.
Delivery notes	These are issued to the buyer by the supplier. They act as a checklist for the buyer to ensure that every item requested on the purchase order has been delivered. The buyer will sign the delivery note when they are satisfied with the delivery.
Timesheets	These are completed by those working on site and are verified by the charge-hand, site manager or supervisor. They detail the start and finish times of each individual working on site. They form the basis of the pay calculation for that worker and the overall time that the job has taken.
Policy documents	These cover health and safety, environmental or customer service issues, among others. They outline the requirements of all those working on the site. They will identify roles and responsibilities, codes of conduct or practice, and methods and remedies for dealing with problems or breaches of policy.

Table 2.11 Documents used in the workplace

Documents for customers and clients

Some documents need to be provided to customers and clients. They are necessary to pass on information and can include records of costs and charges that the customer or client is expected to pay for work carried out. Table 2.12 describes what these documents are and their purpose.

Document	Purpose
Quotations and tenders	Quotations provide written details of the costs of carrying out a particular job. They are based on the specification or requirements of the customer or client. They will usually be written by the main contractor on larger sites. A tender is usually a sealed quotation submitted by a contractor at the same time as tenders from other firms in the hope that their quotation will not only match the requirements but will also be the cheapest and therefore the most likely to win the work.
Estimates	An estimate differs from a quotation because it is not a binding quote but a calculation of the cost based on what the contractor thinks the work may involve.
Invoices	An invoice is a list of materials or services that have been provided. Each has an itemised cost and the total is shown at the bottom of the document, along with any additional charges such as VAT.

Document	Purpose
Account statements	This is a record of all the transactions (invoices and payments) made by a customer or client over a given period. It matches payments by the customer and client against invoices raised by the supplier. It also notes any money still owing or over-payments that may have been made.
Contracts	A contract is a legally binding agreement, usually between a contractor and a customer or client, which states the obligations of both parties. A series of agreements are made as part of the contract. It binds both parties to stick to the agreement, which may detail timescales, level of work or costs.
Contract variations	Contract variations are also legally binding. They may be required if both the supplier and the customer or client agrees to change some of the terms of the original contract. This could mean, for example, additional obligations, renegotiating prices or new timescales.
Handover information	Once a project, such as an installation, has been completed, the installer that commissioned the installation will check that it is performing as expected. Handover information includes: • the commissioning document, which details the performance and the checks or inspections that have been made • an installation certificate, which shows that the work has been carried out in accordance with legal requirements and the manufacturer's recommendations.

Table 2.12 Documents used with customers and clients

Other forms of communication

So far we have mainly focused on written forms of communication.

KEY TERMS

VAT

– Value Added Tax is charged on most goods and services. It is charged by businesses or individuals that have raised invoices in excess of £73,000 per year and is currently 20 per cent of the bill.

ITEM	DESCRIPTION	QUANTITY	UNIT	RATE £	AMOUNT £	
	Superstructure: Suspended upper floor					
A	Supply and fit the following C16 grade preservative treated softwood					
A1	50 × 195 mm joists	250	m	6.44	1610	00
A2	75 × 195 mm joists	60	m	8.58	514	80
A3	38 × 150 mm strutting	70	m	3.85	339	50
	Carried to collection:			£	2464	30

Figure 2.28 An example bill of quantities

However one of the most common forms of communication is the telephone, whether landline or mobile. The key advantage of a conversation is that problems and queries can be immediately sorted out. However the biggest problem is that there is no record of any decisions that have been made. It is therefore often wise to ask for written confirmation of anything that has been agreed, perhaps in the form of an email.

Construction is one of the many industries that still prefer to have hard copies of documents. It has been made much easier to send copies of documents as email attachments. The problem though is having an available printer of sufficient quality and size to print off attached documents.

Performance reviews

As you progress through your construction career you will be expected to attend performance reviews. This is another form of communication between you and your immediate supervisor. Certain levels of performance will be expected and will have been agreed at previous reviews. At each review your performance, compared to those standards, will be examined. It gives both sides an opportunity to look at progress. It can help identify areas where you might need additional training or support. It may also show areas of your work that need improvement and more effort from you.

Meetings

Meetings also offer important opportunities for communication. They are usually quite structured and will have a series of topics that form what is known as an **agenda**.

Meetings should give everyone the opportunity to contribute and make suggestions as to how to go forward on particular projects and deal with problems. Individuals are often given the job of preparing information for meetings and then presenting it for discussion. A disadvantage of meetings is that while they are happening construction work is not taking place. This means that it is important to run meetings efficiently and not waste time – but also to ensure that everything that needs to be discussed is covered so that extra meetings do not have to be arranged.

Letters

Today emails have largely overtaken more traditional forms of communication, but letters can still be important. Letters obviously need to be delivered so take longer to arrive than emails but sometimes things do need to be sent through the post. It is polite to put in a **covering letter** with documents or other written communication with clients.

Signs and posters

On a daily basis you will also see a range of signs and posters around larger construction sites. Signs are used to communicate either warnings or information and a full list of different types of sign, particularly those relating to health and safety, can be seen in Chapter 1. Their purpose is to be clear and informative. Posters are often put up in communal areas, such as where you might have lunch or keep your personal belongings. These are designed to be simple and to give you vital information. One disadvantage of signs and posters is that they are a one-way form of communication so if you need more information about them you will need to speak to your supervisor.

REED TIP

A good supervisor will make sure you understand what is expected of you in terms of quality, quantity, the speed of the work and how you'll be working with other trades.

KEY TERMS

Agenda

– a brief list of topics to be discussed at a meeting, outlining any decisions that need to be made.

Covering letter

– this is a very brief letter, often just one paragraph long, which states the purpose of the communication and lists any other documents that have been included.

TEST YOURSELF

1. If a drawing is at a scale of 1:500, each millimetre in the drawing represents how much on the ground?

   a. 1 m

   b. 500 cm

   c. 500 mm

   d. 500 m

2. What is the other term used to describe an orthographic projection?

   a. First angle

   b. Second angle

   c. Assembly drawing

   d. Isometric

3. Which of the following are examples of a manufacturer's technical information?

   a. Data sheets

   b. User instructions

   c. Catalogues

   d. All of these

4. On a drawing, if you were to see the letters FDN, what would that mean?

   a. The signature of the architect

   b. Foundation Design Network

   c. Foundations

   d. Full distance

5. If a drawing is at a scale of 1:5, how many times smaller is the drawing than the real object?

   a. 5 times

   b. 50 times

   c. Half the size

   d. 500 times

6. Which of the following values is pi?

   a. 3.121

   b. 3.424

   c. 3.142

   d. 3.421

7. Which document is used to give detailed sets of requirements that cover the construction, features, materials and finishes?

   a. Work programme

   b. Purchase order

   c. Policy document

   d. Job specification

8. What is VAT?

   a. Volume Added Turnover

   b. Vehicle Attendance Tax

   c. Voluntary Aided Trading

   d. Value Added Tax

9. Which individual on a typical site would sign off timesheets?

   a. Architect

   b. Site manager/supervisor

   c. Delivery driver

   d. Customer

10. Which are the two main types of communication?

    a. Verbal and written

    b. Telephones and emails

    c. Meetings and memorandum

    d. Plans and faxes

Chapter 3

Unit CSA–L2Core05

UNDERSTANDING CONSTRUCTION TECHNOLOGY

LEARNING OUTCOMES

LO1: Understand the principles of foundation construction

LO2: Understand the principles of floor construction

LO3: Understand the principles of wall construction

LO4: Understand the principles of roof construction

LO5: Understand the supply of utilities and services within construction

LO6: Understand the principle of sustainability within construction

INTRODUCTION

The aim of this chapter is to:

* help you understand the range of building materials used within the construction industry

* help you understand their suitability in the construction of modern buildings.

FOUNDATION CONSTRUCTION

Foundations are the primary element of a building as they support and protect the superstructure (the visible part of the building) above. Foundations are part of the substructure of the building, meaning that they are not visible once the building has been completed.

Foundations spread the load of the superstructure and transfer it to the ground below. They provide the building with structural stability and help to protect the building from any ground movement.

Purpose of foundations

It is important to work out the necessary width of foundations. This depends on the total load of the structure and the load-bearing capacity of the ground or subsoil on which the building is being constructed. This means:

* wide foundations are used when the construction is on weak ground, or the superstructure will be heavy

* narrow foundations are used when the subsoil is capable of carrying a heavy weight, or the building is a relatively light load.

The load that is placed on the foundations spreads into the ground at 45°. **Shear failure** will take place if the thickness of the foundations (T) is less than the projection of the wall or column face on the edge of the foundations (P). This is what leads to subsidence (the ground under the structure sinking or collapsing).

As we will see in this section, the depth of the foundation is dependent on the load-bearing capacity of the subsoil. But for the most part foundations should be 200 mm to 300 mm thick.

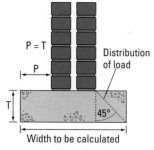

$P = T$
P
Distribution of load
T
45°
Width to be calculated

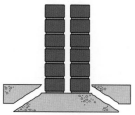

'P' greater than 'T' leads to shear failure

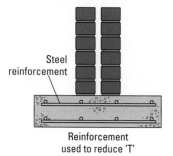

Steel reinforcement

Reinforcement used to reduce 'T'

Figure 3.1 Foundation properties

Different types of foundation

The traditional **strip foundation** is quite narrow and tends to be used for low-rise buildings and dwellings. Most buildings have had unreinforced strip foundations and they were constructed with either brick or block masonry up to the damp course level. Strip foundations can be stepped on sloping ground, in order to cut down on the amount of excavation needed. In poor soil conditions, deep strip foundations can also be used.

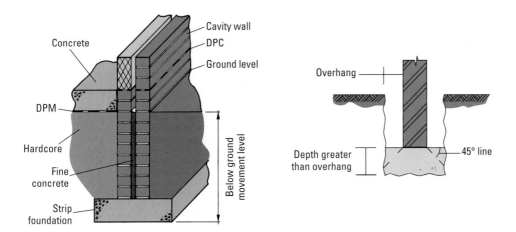

Figure 3.2 Unreinforced strip foundation

Narrow deep strip or trench fill foundations are dug to the foundation depth and then filled with concrete. This reduces excavation, as no bricks or blocks have to be laid into the trench. Trench fill:

* reduces the need to have a wide foundation

* reduces construction time

* speeds up the construction of the footings.

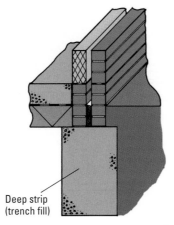

Figure 3.3 Trench fill foundations

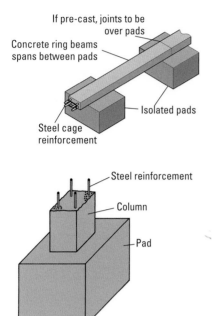

If pre-cast, joints to be over pads

Concrete ring beams spans between pads

Isolated pads

Steel cage reinforcement

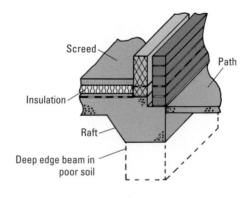

Steel reinforcement

Column

Pad

Figure 3.4 Pad foundations

Pad foundations tend to be used for structures that have either a concrete or a steel frame. The pads are placed to support the columns, which transfer the load of the building into the subsoil.

Pile foundations tend to be used for high-rise buildings or where the subsoil is unstable. Holes are bored into the ground and filled with concrete or pre-cast concrete, steel or timber posts are driven into the ground. These piles are then spanned with concrete ring beams with steel reinforcement so that the load of the building is transferred deeper into the ground below. Pile foundations can be short or long depending on how high the building is or how bad the soil conditions are.

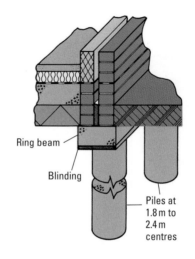

Ring beam

Blinding

Piles at 1.8 m to 2.4 m centres

Figure 3.5 Pile foundations

Raft foundations are used when there is a danger that the subsoil is unstable. A large concrete slab reinforced with steel bars is used to outline the whole footprint of the building. It has an edge beam to take the load from the walls, which is transferred over the whole raft. This means that the building effectively 'floats' on the ground surface on top of the concrete raft.

Screed

Path

Insulation

Raft

Deep edge beam in poor soil

Figure 3.6 Raft foundations

Selecting a foundation

One of the first things that a structural engineer will look at when they investigate a site is the nature of the soil and issues such as the water table (where groundwater begins). The type of soil or ground conditions are be very important, as is the possibility of ground movement.

Table 3.1 shows different types of subsoil and how they can affect the choice of foundation.

Subsoil type	Characteristics
Rock	High load bearing but there may be cracks or faults in the rock, which could collapse.
Granular	Medium to high load bearing and can be compacted sand or gravel. If there is a danger of flooding the sand can be washed away.
Cohesive	Low to medium load bearing, such as clay and silt. These are relatively stable, but may have problems with water.
Organic	Low load bearing, such as peat and topsoil. Organic material must be removed before starting the foundations. There is also a great deal of air and water present in the soil.

Table 3.1 Different types of subsoil

The ground may move, particularly if the conditions are wet, extremely dry or there are extremes of temperature. Clay, for example, will shrink in the hot summer months and swell up again in the wet winter months. Frost can affect the water in the ground, causing it to expand.

Ground movement is also affected by the proximity of trees and large shrubs. They will absorb water from the soil, which can dry out the subsoil. This causes the soil underneath the foundations to collapse.

The end use of the building

The other key factor when selecting a foundation is the end use of the building:

* Strip foundation – this is the most common and cheapest type of foundation. Strip foundations are used for low to medium rise domestic and industrial buildings, as the load-bearing will not be high.

* Raft foundation – this is only ever really used when the ground on which the building is being constructed is very soft. It is also sometimes used when the ground across the area is likely to react in different ways because of the weight of the building. In areas of the UK where there has been mining, for example, raft foundations are quite common, as the building could subside. The raft is a rigid, concrete slab reinforced with steel bars. The load of the building is spread across the whole area of the raft.

* Piled foundations – these are used for high-rise buildings where the building will have a high load or where the soil is found to be poor.

Materials used in the construction of foundations

Concrete

Concrete is used to produce a strong and durable foundation. The concrete needs to be poured into the foundation with some care. The size of the foundation will usually determine whether the concrete is actually mixed on site or brought in, in a ready-mixed state, from a supplier. For smaller foundations a concrete mixer and wheelbarrows are usually sufficient. The concrete is then poured into the foundation using a chute.

Concrete consists of both fine and coarse aggregate, along with water, cement and additives if required.

Aggregates

Aggregates are basically fillers. The coarse aggregate is usually either crushed rock or gravel. The grains are 5mm or larger.

Fine aggregate is usually sand that has grains smaller than 5mm.

The fine aggregate fills up any gaps between the particles in the coarse aggregate.

Cement

Cement is an adhesive or binder. It is Portland stone, crushed, burnt and crushed again and mixed with limestone. The materials are powdered and then mixed together to create a fine powder, which is then fired in a kiln.

Water

Potable water, which is water that is suitable for drinking, should be used when making concrete. The reason for this is that drinkable water has not been contaminated and it does not have organic material in it that could rot and cause the concrete to crack. The water mixes with the cement and then coats the aggregate. This effectively bonds everything together.

Additives

Additives, or admixtures, make it possible to control the setting time and other aspects of fresh concrete. It allows you to have greater control over the concrete. Common add mixtures can accelerate the setting time, or reduce the amount of water required. They can:

* give you higher strength concrete
* provide protection against corrosion
* accelerate the time the concrete needs to set
* reduce the speed at which the concrete sets
* provide protection against cracking as the concrete sets (prevent shrinkage)
* improve the flow of the concrete

* improve the finish of the concrete

* provide hot or cold weather protection (a drop or rise in temperature can change the amount of time that a concrete needs to set, so these add mixtures compensate for that).

Reinforcement

Steel bars or mesh can be used to give the foundation additional strength and support. It can also stop the foundation from cracking. Concrete is very good at dealing with loads, so weight coming from above is something concrete can deal with. But when concrete foundations are wide, and parts of them are under tension, there is a danger it may crack.

Concrete should also be levelled, usually with a vibrator or a compactor, although newer types of concrete are self-compacting. All concrete needs to be laid on well-compacted ground.

Figure 3.7 Reinforcement using steel bars (or mesh)

FLOOR CONSTRUCTION

For most domestic buildings floorboards or sheets are laid over timber joists. In other cases, and in most industrial buildings, the ground floors have a block and beam construction with hard core. They then have a damp-proof membrane and over the top is solid concrete.

A floor is a level surface that provides some insulation and carries any loads (for example furniture) and to transfer those loads.

The ground floors have additional purposes. They must stop moisture from entering the building from the ground. They also need to prevent plant or tree roots from entering the building.

Ground floors

For ground floors there are two options:

* Solid – is in contact with the ground.

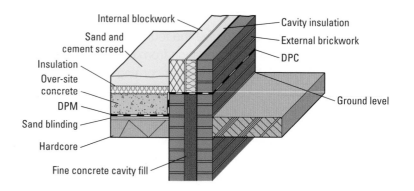

Figure 3.8 Solid ground floors

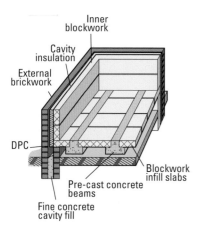

Figure 3.9 Suspended ground floors

* Suspended – the floor does not touch the ground and spans between walls in the building. Effectively there is a void beneath the floor, with air bricks in external walls to allow for ventilation.

The options for ground floors are more complicated than those for upper floors. This is because the ground floors need to perform several functions. It is quite rare for modern buildings to have timber joists and floorboards. Suspended ground floors and traditional timber floors tend to be seen in older buildings. It is far more common to have solid ground floors, or to have timber floors over concrete floors, which are known as floating ground floors.

The key options are outlined in Table 3.2.

Type of floor	Construction and characteristics
Solid	One construction method is to use hard core as the base, with a layer of sand and a layer of insulation such as Celotex, usually 100 mm thick, and then covered with a damp-proof membrane. The concrete is then poured into the foundation. To provide a smooth finish for floor finishes a cement and sand screed is applied, usually after the building has been made watertight..
Timber suspended	A similar process to a solid ground floor is carried out but then, on top of this, dwarf or sleeper walls are built. These are used to support the timber floor. Air bricks are also added to provide necessary ventilation. Joists are then spaced out along the dwarf walls. A damp-proof course is inserted under the floor joists and then floorboards or sheets placed on top of the joists.
Beam and block suspended	Concrete beams and lightweight concrete slabs or blocks are used to create the basic flooring. The beams are evenly spaced across the foundation and gaps between the beams are filled with blocks to form the floor. The blocks and beams are then insulated and it is finished off with either a cement screed or a timber floating floor.
Floating	This timber construction goes over the top of concrete floors. Bearers are put down and then the boarding or sheets are fixed to the bearers. The weight of the boards themselves hold them in place.

Table 3.2 Construction of ground floors

Upper floors

Timber is usually used for these suspended floors in homes and other types of dwellings. In industrial buildings concrete tends to be used.

Timber suspended upper floor

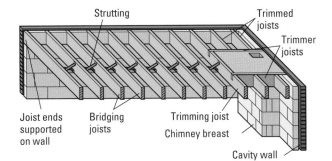

Strutting

Trimmed joists

Trimmer joists

Joist ends supported on wall

Bridging joists

Trimming joist

Chimney breast

Cavity wall

Concrete suspended upper floor

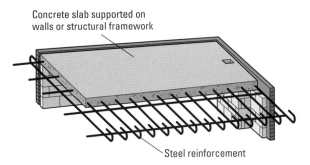

Concrete slab supported on walls or structural framework

Steel reinforcement

Figure 3.10 Upper floors

For dwellings, bridging joists are used. These are supported at their ends by load-bearing walls. Boarding or sheets provide the flooring for the room on the top of the joists. Underneath the joists plasterboard creates the basis of the ceiling for the room below.

It is also possible to fill the voids between the floorboards and the plasterboard with insulation. Insulation not only helps to prevent heat loss, but can also reduce noise.

Concrete suspended floors are usually either cast on site or available as ready-cast units. They are effectively locked into the structure of the building by steel reinforcement. If the concrete floors are being cast on site then **formwork** is needed. Concrete floors tend to be used in many modern buildings, particularly industrial ones, as they offer greater load bearing capacity, have greater fire resistance and are more sound resistant.

WALL CONSTRUCTION

Walls have a number of different purposes:

* They hold up the roof.

* They provide protection against the elements.

* They keep the occupants of the building warm.

Many buildings now have double walls, which means as follows:

* The outside wall is a wet one because it is exposed to the elements outside the building.

* The internal wall is dry but it needs to be kept separate from the outside wall by a cavity.

* The cavity or gap acts as a barrier against damp and also provides some heat insulation.

* The cavity can be completely filled or part-filled depending on the insulation value required by Building Regulations.

Within the building there are other walls. These internal walls divide up the space within the building. These do not have to cope with all of the demands of the external walls. As a result, they do not necessarily have to be insulated and are, therefore, thinner. They are block and then covered with plaster. Alternatively they can be a timber framework, which is also known as stud work, and again can be covered with plasterboard.

Different types of wall construction and structural considerations

In addition to walls being external or internal, they can also be classed as being load bearing or non-load bearing.

Internal walls can be either load bearing or non-load bearing. In both internal and external walls, where they are load bearing, any gaps or openings for windows or doors have to be bridged. This is achieved by using either arches or lintels. These support the weight of the wall above the opening.

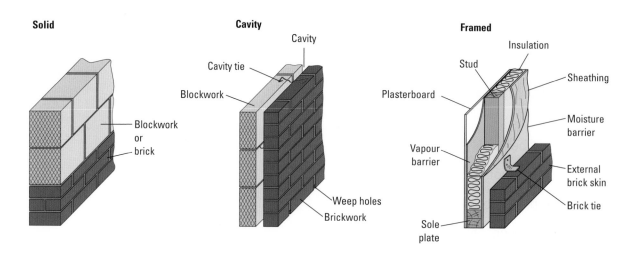

Figure 3.11 Some examples of external wall construction

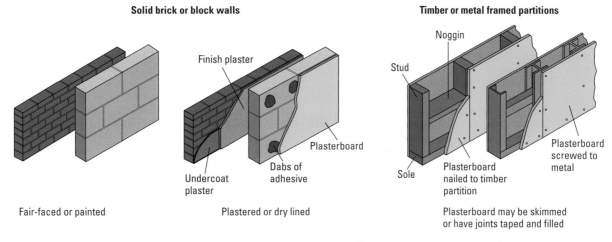

Solid brick or block walls

Fair-faced or painted

Finish plaster

Plasterboard

Dabs of adhesive

Undercoat plaster

Plastered or dry lined

Timber or metal framed partitions

Noggin

Stud

Sole

Plasterboard nailed to timber partition

Plasterboard screwed to metal

Plasterboard may be skimmed or have joints taped and filled

Figure 3.12 Some examples of internal wall construction

Solid masonry

In modern builds solid masonry is quite rare, as it uses up a lot of bricks and blocks. External solid walls tend to be much thinner and made from lightweight blocks in modern builds. They will have some kind of waterproof surface over the top of them, which can be made of render, plastic, metal or timber.

Cavity masonry

As we have seen, cavity walls have an outer and an inner wall and a cavity between them. Usually solid walling, or blockwork, is built up to ground level and then the cavity walling continues to the full height of the building. These cavity walls are ideal for most buildings up to medium height.

Many industrial buildings have cavity walls for the lower part of the building and then have insulated steel panels for the top part of the building.

The usual technique is to have brick for the outer wall and an insulating block for the inner wall. The gap or cavity can then be filled with an insulation material.

Timber framed

Panels made of timber, or in some cases steel, are used to construct walls. They can either be load bearing or non-load bearing and can also be used for the outside of the building or for internal walling. The panels are solid structures and the spaces between the vertical struts (studs) and the horizontal struts (head or sole plates) can be filled with insulation material.

Internal walls or partitions

Internal walls tend to be either solid or framed. Solid walls are made up from blocks. In many industrial buildings the blocks are actually exposed and can be left in their natural state or painted. In domestic buildings plasterboard is usually bonded to the surface and then plastered over to provide a smoother finish.

It is more common for domestic buildings to have framed internal walls, which are known as stud partitions. These are exactly the same as other framed walling, but will usually have plasterboard fixed to them. They would then receive a skimmed coat of plaster to provide the smooth finish.

Damp-proof membrane (DPM) and damp-proof course (DPC)

Damp-proof membranes are installed under the concrete in ground floors in order to ensure that ground moisture does not enter the building. Effectively the membrane waterproofs the building.

Damp-proof courses are a continuation of the damp-proof membrane. They are built into a horizontal course of either block and brickwork, which is a minimum of 150 mm above the exterior ground level. DPCs are also designed to stop moisture from coming up from the ground, entering the wall and then getting into the building. The most common DPC is a polythene sheet called visqueen DPC. It comes in rolls to the appropriate width for the wall.

In older buildings lead, bitumen or slate would have been used as a DPC.

KEY TERMS

Perpendicular

– this means at right angles or 90° to the horizontal.

ROOF CONSTRUCTION

In a country such as the UK, with a great deal of rain and sometimes snowy weather, it makes sense for roofs to be pitched. Pitched means built at an angle. The idea is that the rain and snow falls down the angle and off the edge of the roof or into gutters rather than lying on the roof.

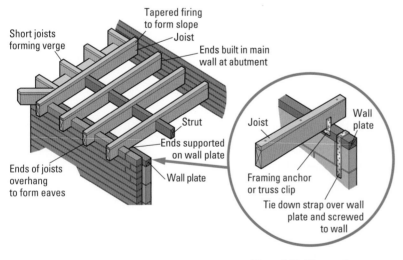

Figure 3.13 Flat roof structure

However, not all roofs are pitched. In fact many domestic dwelling extensions have flat roofs. A great number of industrial buildings have entirely flat roofs. The problem with a flat roof is that it needs to be able to support itself, but just as importantly it needs to be able to carry the additional weight of snow or rain. This means that large flat roofs may have to have steel sections (known as trusses) or even reinforced concrete and beams to increase their load-bearing capacity.

Roofs also provide stability to the walls by tying them together. As we will see, there are several different types of roof. These are usually identified by their pitch or shape.

Types of roof construction

The roof is made up of the rafters and beams. Everything above the framework is regarded as a roof covering, such as slates, tiles and felt.

Table 3.3 outlines some of the key characteristics of different types of roof.

Roof type	Characteristics	How it looks
Flat	This is a roof that has a slope of less than 10°. Generally flat roofs are used for smaller extensions to dwellings and on garages. Traditionally they would have had bitumen felt, although it is becoming more common for fibreglass to be used.	 Figure 3.14
Mono-pitch	This is a roof that has a single sloping surface but is not fixed to another building or wall. The front and back walls could be different heights, or the other exposed surface of the roof is perpendicular.	 Figure 3.15
Gambrel roof	This is a roof that has two differently angled slopes. Usually the upper part of the roof has a fairly shallow pitch or slope and the lower part of the roof has a steeper slope.	 Figure 3.16
Couple roof	This is often called gable end and is one of the most common types of roof for dwellings. A gable is a wall with a triangular upper part. This supports the roof in construction using purlins. This means that the roof has two sloping surfaces, which come down from the ridge to the eaves.	 Figure 3.17
Hipped roof	Hipped roofs have slopes on three or four four sides. There are also hipped roofs with single, straight gables.	 Figure 3.18
Lean-to	A lean to is similar to a mono-pitched roof except it is abutted to a wall. The slope is greater than 10°. The higher part of the roof is fixed to a higher wall.	 Figure 3.19

Table 3.3 Different types of roof

Roofing components

Each visible part of a roof has a specific name and purpose. Table 3.4 explains each of these individual features.

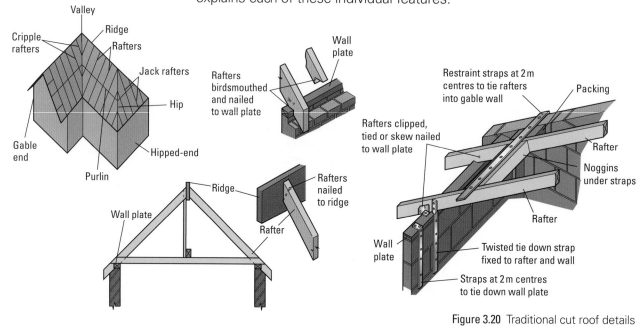

Figure 3.20 Traditional cut roof details

Roof feature	Description
Ridge	This is the top of the roof and the junction of the sloping sides. It is the peak, where the rafters meet.
Purlin	This is a beam that supports the mid-span section of rafters.
Firings	These are angled pieces of timber that are placed on the rafters to create a slope.
Batten	Roof battens are thin strips, usually of wood, which provide a fixing point for either roofing sheets or roof tiles.
Tile	These can be made from clay, slate, concrete or plastic. They are placed in regular, overlapping rows and fixed to the battens.
Fascia	This is a horizontal, decorative board. It is usually a wooden board, although it can be PVC. It is fixed to the ends of the rafters at eaves level and is both a decorative feature and a fixing for rainwater goods.
Wall plate	This is a horizontal timber that is placed at the top of a wall at eaves level. It holds the ends of joists or rafters.
Bracings	Roof rafters need to be braced to make them more rigid and stable. These bracings prevent the roof from buckling. Usually there are several braces in a typical roof.
Felt	Roofing felt has two elements – it has a waterproofing agent (bitumen) and what is known as a carrier. The carrier can be either a polyester sheet or a glass fibre sheet. Roofing felt tends to be used for flat roofs and for roofs with a shallow pitch.
Slate	Slate roofing tiles are usually fixed to timber battens with double nails. They have a lifespan of between 80 and 100 years.
Flashings	Wherever there is a joint or angle on a roof, a thin sheet of either lead or another waterproof material is added. In the past this tended always to be lead. Many different types of flashing can now be used but all have the role of preventing water penetrating into joints, such as on abutments to walls and around the chimney stack.
Rafter	Roof rafters are the main structural components of the roof. They are the framework. They rest on supporting walls. The rafters are set at an angle on sloped roofs or horizontal on a flat roof.
Apex	The apex is the highest point of the roof, usually the ridge line.

Roof feature	Description
Soffit	Soffits are the lower part, or overhanging part, of the eaves. In other words they are the underside of the eaves. A flat section of timber or plastic is usually fixed to the soffit to ensure water tightness.
Bargeboard	This is an ornamental feature, which is fixed to the gable end of a roof in order to hide the ends of roof timbers.
Eaves	These are the area found at the foot of the rafter. They are not always visible as they can be flush. In modern construction, the eaves have two parts: the visible eaves projection and the hidden eaves projection.

Table 3.4 Parts of a roof

Roof coverings

There are many different types of materials that can be used to cover the roof. Even tiles and slates come in a wide variety of shapes and sizes, along with colours and different finishes.

In many cases the type of roof covering is determined by the traditional and local styles in the area. Local authorities want roof coverings that are not too far from the common style in the area. This does not stop manufacturers from coming up with new ideas, however, which can add benefits during construction. There is much innovation and labour saving that also helps to minimise build costs.

Affordable clay tiles, for example, make it possible to use traditional materials that had been out of the budget of many construction jobs for a number of years.

The Table 3.5 outlines some of the more common types of roof covering and describes their main characteristics and use.

Roof covering	Description
Felt Figure 3.21	Felt is used as a waterproof barrier. Internal felt is rolled over the top of the rafters. The strips are overlapped to provide a permanent waterproof barrier. They are then battened down and another roof covering, such as slate or tile, placed over the top of them. For flat roofs, felt is used as the external roof covering and is covered in a waterproof material, such as bitumen.
Slate Figure 3.22	Slate is a flat, natural substance, which is laid onto the battens with each slate tile overlapping the top of the slate in the row directly below it. The slate tiles are either nailed or hooked into place.
Tile Figure 3.23	There is a huge variety of roofing tiles, made from clay, ceramics or concrete. They are designed and moulded so that they overlap with one another and are fixed to the roof in a similar way to slate tiles.

Roof covering	Description
Metals  Figure 3.24	There are many different types of metal roof covering, such as corrugated sheets, flat sheets, box profile sheets or even sheets that have a tile effect. The metal is galvanised and plastic coated to provide a durable and long-lasting waterproof surface.

Table 3.5 Roof coverings

CASE STUDY

South Tyneside Homes

South Tyneside Council's Housing Company

How to impress in interviews

Andrea Dickson and Gillian Jenkins sit on the interview panels for apprenticeship applications at South Tyneside Homes.

'Interviews are all about the three Ps: Preparation, Presentation and Personality.

An applicant should turn up with some knowledge about the apprenticeship programme and the company itself. For example, knowing how long it is, that they have to go to college and to work – and don't say, "I was hoping you'd tell me about it"! If they've done a bit of research, it will show through and work in their favour – especially if they can explain why it is that they want to work here.

It sets them up for the interview if they come in smartly dressed. We're not marking them on that, but it does show respect for the situation. It's still a formal process and although we try to make them feel at ease as much as we possibly can, there's no getting away from the fact that they're applying for a job and it is a formal setting.

The interviews are a chance to tell the company about themselves: what they do in their spare time, what their greatest achievements have been and why. Applicants should talk about what interests them, for example, are they really interested in becoming a joiner or is that something their parents want them to do? An apprenticeship has to be something they really want to do – if they have enthusiasm for the programme, then they'll fly through it. If not, it's a very long three to four years. Without that passion for it, the whole process will be a struggle; they'll come in late to work and might even fail exams.

We also talk to them about any customer service experiences they've had, working in a team, project working (for example, a time you had to complete a task and what steps you took), as well as asking some questions about health and safety awareness.'

SUPPLY OF UTILITIES AND SERVICES

Most but not all dwellings and other structures are connected in some way to a wide range of utilities and services. In the majority of cities, towns and villages structures are connected to key utilities and services, such as a sewer system, potable (drinking) water, gas and electricity. This is not always the case for more remote structures.

Whenever construction work is carried out, whether it is on an existing structure or a new build, the supply of utilities and services or the linking up of these parts of the **infrastructure** are very important. Often they will require the services of specialist engineers from the **service provider**.

Table 3.6 outlines the main utilities and services that are provided to most structures.

KEY TERMS

Infrastructure

– these are basic facilities, such as a power supply, a road network and a communication link.

Service provider

– these are companies or organisations that provide utilities, such as gas, water, communications or electricity.

Utility or service	Description
Drainage	Drainage is delivered by a range of water and sewerage companies. They are responsible for ensuring that surface water can drain away into their system.
Waste water and sewerage	Any waste water and sewage generated by the occupants of a structure needs to have the necessary pipework to link it to the main sewerage system. It is then sent to a sewage treatment works via the pipework. Remote areas may not be connected to the sewerage system so use septic tanks and cesspools.
Water	Each structure should be linked to the water supply that provides wholesome, potable drinking water. The pipework linking the structure to the water supply needs to be protected to ensure that backflow from any other source does not contaminate the system.
Gas	Each area has a range of different gas suppliers. This is delivered via a service pipe from the main system into the structure. Areas that do not have access to the main gas supply system use gas contained in bottles.
Electricity	The National Grid provides electricity to a variety of different electricity suppliers. It is the National Grid that operates and maintains the cabling. There are around 28 million individual customers in the UK.
Communications (telephone, data, cable)	There are several ways in which telecommunications can be linked to a structure. Traditional telephone poles hold up copper cables and not only provide telephone but also internet access to structures. In cities and many of the larger towns this system is being replaced by cables that are fibre optic and run underground. These are then linked to each individual structure.
Ducting (heating and ventilation)	Heating and ventilation engineers install and maintain duct work. The complex systems are known as **HVAC**. These systems can transfer air for heating or cooling of the structure. The overall system can also provide hot and cold water systems, along with ventilation.

Table 3.6 Services and utilities

SUSTAINABILITY AND INCORPORATING SUSTAINABILITY INTO CONSTRUCTION PROJECTS

Sustainability is something that we all need to be concerned about as the earth's resources are used up rapidly and climate change becomes an ever-bigger issue. Carbon is present in all fossil fuels, such as coal or natural gas. Burning fossil fuels releases carbon dioxide, which is a greenhouse gas linked to climate change.

Energy conservation aims to reduce the amount of carbon dioxide in the atmosphere. The idea is to do this by making buildings better insulated and, at the same time, make heating appliances more efficient. Sustainability also means attempting to generate energy using renewable and/or low or zero carbon methods.

According to the government's Environment Agency, sustainable construction means using resources in the most efficient way. It also means cutting down on waste on site and reducing the amount of materials that have to be disposed of and put into **landfill**.

In order to achieve sustainable construction the Environment Agency recommends

* reducing construction, demolition and excavation waste that needs to go to landfill

* cutting back on carbon emissions from construction transport and machinery

* responsibly sourcing materials

* cutting back on the amount of water that is wasted

* making sure construction does not have a negative impact on **biodiversity**.

Sustainable construction and incorporating it into construction projects

In the past buildings were generally constructed as quickly as possible and at the lowest cost. More recently the idea of sustainable construction focuses on ensuring that the building is not only of good quality and that it is affordable, but that it is also efficient in terms of energy use and resources.

Sustainable construction also means having the least negative environmental impact. So this means minimising the use of raw materials, energy, land and water. This is not only during the build but also for the lifetime of the building.

Finite and renewable resources

We all know that resources such as coal and oil will eventually run out. These are examples of finite resources.

Oil, however, is not just used as fuel – it is in plastic, dyes, lubricants and textiles. All of these are used in the construction process.

Renewable resources are those that are produced either by moving water, the sun or the wind. Materials that come from plants, such as biodiesel, or the oils used to make adhesives, are all examples of renewable resources.

Figure 3.25 Most modern new-builds follow sustainable principles

The construction process itself is only part of the problem. It is important to consider the longer-term impact and demands that the building will have on the environment. This is why there has been a drive towards sustainable homes and there is a Code for Sustainable Homes.

Construction and the environment

In 2010, construction, demolition and excavation produced 20 million tonnes of waste that had to go into landfill. The construction industry is also responsible for most illegal fly tipping (illegally dumping waste). In any year the Environment Agency responds to around 350 pollution incidents caused as a result of construction.

Regardless of the size of the construction job, everyone in construction is responsible for the impact they have on the environment. Good site layout, planning and management can help to reduce these problems.

Sustainable construction helps to encourage this because it means managing resources in a more efficient way, reducing waste, recycling where possible and reducing your **carbon footprint**.

Architecture and design

The Code for Sustainable Homes Rating Scheme was introduced in 2007. Many local authorities have instructed their planning departments to encourage sustainable development. This begins with the work of the architect who designs the building.

DID YOU KNOW?

Search on the internet for 'sustainable building' and 'improving energy efficiency' to find out more about the latest technologies and products.

KEY TERMS

Carbon footprint

– This is the amount of carbon dioxide produced by a project. This not only includes burning carbon-based fuels such as petrol, gas, oil or coal, but includes the carbon that is generated in the production of materials and equipment.

Local authorities ask that architects and building designers:

* ensure the land is safe for development – so if it is contaminated this is dealt with first

* ensure access to and protect the natural environment – this supports biodiversity and tries to create open spaces for local people

* reduce the negative impact on the local environment – buildings should keep noise, air, light and water pollution down to a minimum

* conserve natural resources and cut back carbon emissions – this covers energy, materials and water

* ensure comfort and security – good access, close to public transport, safe parking and protection against flooding.

Figure 3.26 Sustainable developments aim to be pleasant places to live

Using locally managed resources

The construction industry imports nearly 6 million cubic metres of sawn wood each year. However there is plenty of scope to use the many millions of cubic metres of timber produced in managed forests, particularly in Scotland.

Local timber can be used for a wide variety of different construction projects:

* Softwood – including pines, firs, larch and spruce – for panels, decking, fencing and internal flooring.

* Hardwood – including oak, chestnut, ash, beech and sycamore – for a wide variety of internal joinery.

Eco-friendly, sustainable manufactured products and environmentally resourced timber

There are now many suppliers that offer sustainable building materials as a green alternative. Tiles, for example, are now made from recycled plastic bottles and stone particles.

There is a National Green Specification database of all environmentally friendly building materials. This provides a checklist where it is possible to compare specifications of sustainable products to traditionally manufactured products, such as bricks.

Simple changes can be made, such as using timber or ethylene-based plastics instead of UPVC window frames, to ensure a building uses more sustainable materials.

As we have seen, finding locally managed resources such as timber makes sense in terms of cost and in terms of protecting the environment. There are many alternatives to traditional resources that could help protect the environment.

The Timber Trade Federation produces a Timber Certification System. This ensures that wood products are labelled to show that they are produced in sustainable forests.

Around 80 per cent of all the softwood used in construction comes from Scandinavia or Russia. Another 15 per cent comes from the rest of Europe, or even North America. The remaining 5 per cent comes from tropical countries, and is usually sourced from sustainable forests.

Alternative methods of building

The most common type of construction is, of course, brick and block work. However there are plenty of other options:

* Timber frame

* Insulated concrete formwork – where a polystyrene mould is filled with reinforced concrete.

* Structural insulated panels – where buildings are made up of rigid building boards, rather like huge sandwiches.

* Modular construction – this uses similar materials and techniques to standard construction, but the units are built off site and transported ready-constructed to their location.

Figure 3.27 Window frames made from timber

DID YOU KNOW?

www.recycledproducts.org.uk has a long list of recycled surfacing products, such as tiles, recycled wood and paving and details of local suppliers.

Figure 3.28 Timber Certification System

Figure 3.29 Green roofing

Figure 3.30 Flooring made from cork

Alternatives to roofing and flooring

There are alternatives to traditional flooring and roofing, all of which are greener and more sustainable. Green roofing has become an increasing trend in recent years. Metal roofs made of steel, aluminium or copper are lightweight and often use a high percentage of recycled metal. Solar roof shingles, or solar roof laminates, while expensive, help to reduce the use of electricity and heating of the dwelling. Some buildings even have a green roof, which consists of a waterproof membrane, a growing medium and plants such as grass or sedum.

Just as roofs are becoming greener, so too are the options for flooring. The use of bamboo, eucalyptus or cork is becoming more common. A new version of linoleum has been developed with **biodegradable**, **organic** ingredients. Some buildings are also using concrete rather than traditional timber floorboards and joists. The concrete can be coloured, stained or patterned.

An increasing trend has been for what is known as off-site manufacture (OSM). European businesses, particularly those in Germany, have built over 100,000 houses. The entire house is manufactured in a factory and then assembled on site. Walls, floors, roofs, windows and doors with built-in electrics and plumbing, all arrive on a lorry. Some manufacturers even offer completely finished dwellings, including carpets and curtains. Many of these modular buildings are actually designed to be far more energy efficient than traditional brick and block constructions. Many come ready fitted with heat pumps, solar panels and triple-glazed windows.

KEY TERMS

Biodegradable

– the material will more easily break down when it is no longer needed. This breaking down process is done by micro-organisms.

Organic

– this is a natural substance, usually extracted from plants.

Energy efficiency and incorporating it into construction projects

Energy efficiency involves using less energy to provide the same level of output. The plan is to try to cut the world's energy needs by 30 per cent before 2050. This means producing more energy efficient buildings. It also means using energy efficient methods to produce materials and resources needed to construct buildings.

Building Regulations

In terms of energy conservation, the most important UK law is the Building Regulations 2010, particularly Part L. The Building Regulations:

* list the minimum efficiency requirements

* provide guidance on compliance, the main testing methods, installation and control

* cover both new dwellings and existing dwellings.

A key part of the regulations is the Standard Assessment Procedure (SAP), which measures or estimates the energy efficiency performance of buildings.

Local planning authorities also now require that all new developments generate at least 10 per cent of their energy from renewable sources. This means that each new project has to be assessed one at a time.

Energy conservation

By law, each local authority is required to reduce carbon dioxide emissions and to encourage the conservation of energy. This means that everyone has a responsibility in some way to conserve energy:

* Clients, along with building designers, are required to include energy efficient technology in the build.

* Contractors and sub-contractors have to follow these design guidelines. They also need to play a role in conserving energy and resources when actually working on site.

* Suppliers of products are required by law to provide information on energy consumption.

In addition, new energy efficiency schemes and building regulations cover the energy performance of buildings. Each new build is required to have an Energy Performance Certificate. This rates a building's energy efficiency from A (which is very efficient) to G (which is least efficient).

Some building designers have also begun to adopt other voluntary ways of attempting to protect the environment. These include BREEAM, which is an environmental assessment method, and the Code for Sustainable Homes, which is a certification of sustainability for new builds.

energy®
saving
trust

Figure 3.31 The Energy Saving Trust encourages builders to use less wasteful building techniques and more energy efficient construction

DID YOU KNOW?

The Energy Saving Trust has lots of information about construction. Its website is www.energysavingtrust.org.uk.

High, low and zero carbon

When we look at energy sources, we consider their environmental impact in terms of how much carbon dioxide they release. Accordingly, energy sources can be split into three different groups:

- high carbon – those that release a lot of carbon dioxide
- low carbon – those that release some carbon dioxide
- zero carbon – those that do not release any carbon dioxide.

Some examples of high carbon, low carbon and zero carbon energy sources are given in Table 3.7 below.

High carbon energy source	Description
Natural gas or LPG	Piped natural gas or liquid petroleum gas stored in bottles
Fuel oils	Domestic fuel oil, such as diesel
Solid fuels	Coal, coke and peat
Electricity	Generated from non-renewable sources, such as coal-fired power stations
Low carbon energy source	
Solar thermal	Panels used to capture energy from the sun to heat water
Solid fuel	Biomass such as logs, wood chips and pellets
Hydrogen fuel cells	Converts chemical energy into electrical energy
Heat pumps	Devices that convert low temperature heat into higher temperature heat
Combined heat and power (CHP)	Generates electricity as well as heat for water and space heating
Combined cooling, heat and power (CCHP)	A variation on CHP that also provides a basic air conditioning system
Zero carbon energy	
Electricity/wind	Uses natural wind resources to generate electrical energy
Electricity/tidal	Uses wave power to generate electrical energy
Hydroelectric	Uses the natural flow of rivers and streams to generate electrical energy
Solar photovoltaic	Uses solar cells to convert light energy from the sun into electricity

Table 3.7 High, low and zero carbon energy sources

It is important to try to conserve non-renewable energy so that there will be sufficient fuel for the future. The fuel has to last as long as is necessary to completely replace it with renewable sources, such as wind or solar energy.

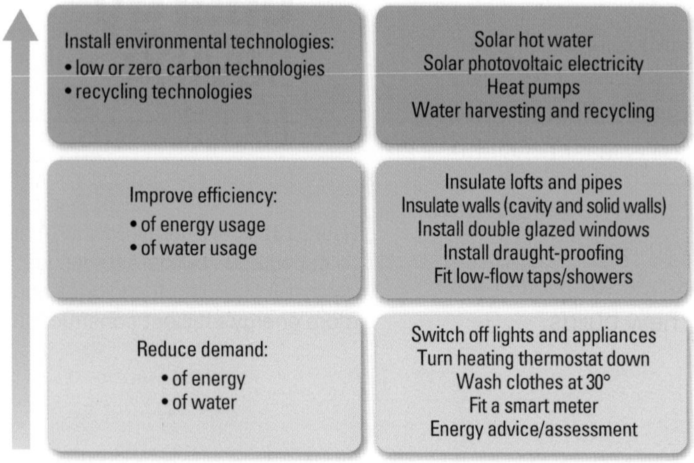
Install environmental technologies:
• low or zero carbon technologies
• recycling technologies
Solar hot water
Solar photovoltaic electricity
Heat pumps
Water harvesting and recycling
Improve efficiency:
• of energy usage
• of water usage
Insulate lofts and pipes
Insulate walls (cavity and solid walls)
Install double glazed windows
Install draught-proofing
Fit low-flow taps/showers
Reduce demand:
• of energy
• of water
Switch off lights and appliances
Turn heating thermostat down
Wash clothes at 30°
Fit a smart meter
Energy advice/assessment

Figure 3.32 Working towards reducing carbon emissions

Alternative energy sources

There are several new ways in which we can harness the power of water, the sun and the wind to provide us with new heating sources. All of these systems are considered to be far more energy efficient than traditional heating systems, which rely on gas, oil, electricity or other fossil fuels.

Solar thermal

At the heart of this system is the solar collector, which is often referred to as a solar panel. The idea is that the collector absorbs energy from the sun, which is then converted into heat. This heat is then applied to the system's heat transfer fluid.

The system uses a differential temperature controller (DTC) that controls the system's circulating pump when solar energy is available and there is a demand for water to be heated.

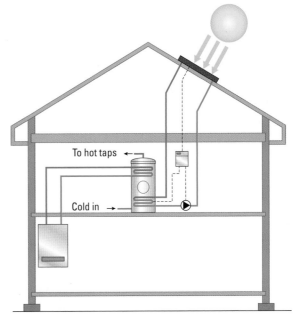

Figure 3.33 Solar thermal hot water system

In the UK, due to the lack of guaranteed solar energy, solar thermal hot water systems often have an auxiliary heat source, such as an immersion heater.

Biomass (solid fuel)

Biomass stoves burn either pellets or logs. Some have integrated hoppers that transfer pellets to the burner. Biomass boilers are available for pellets, woodchips or logs. Most of them have automated systems to clean the heat exchanger surfaces. They can provide heat for domestic hot water and space heating.

Stove providing
room heat only

Stove providing
room heat and
domestic hot water

Stove providing
room heat, domestic
hot water and heating

Figure 3.34 Biomass stoves output options

Heat pumps

Heat pumps convert low temperature heat from air, ground or water sources to higher temperature heat. They can be used in ducted air or piped water **heat sink** systems.

> **KEY TERMS**
>
> **Heat sink**
>
> – this is a heat exchanger that transfers heat from one source into a fluid, such as in refrigeration, air-conditioning or the radiator in a car.

Geothermal

– relating to the internal heat energy of the earth.

There are different arrangements for each of the three main systems:

* Air source pumps operate at temperatures down to minus 20°C.

* Ground source pumps operate on **geothermal** ground heat.

* Water source systems can be used where there is a suitable water source, such as a pond or lake.

The heat pump system's efficiency relies on the temperature difference between the heat source and the heat sink.

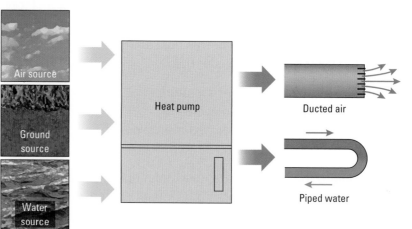

Figure 3.35 Heat pump input and output options

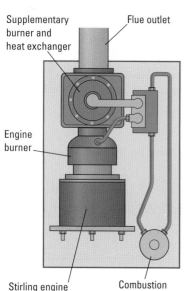

Figure 3.36 Example of a MCHP (micro combined heat and power) unit

Combined heat and power (CHP) and combined cooling heat and power (CCHP) units

These are similar to heating system boilers, but they generate electricity as well as heat for hot water or space heating (or cooling). Electricity is generated along with sufficient energy to heat water and to provide space heating.

Wind turbines

Freestanding or building-mounted wind turbines capture the energy from wind to generate electrical energy. The wind passes across rotor blades of a turbine, which causes the hub to turn. The hub is connected by a shaft to a gearbox. This increases the speed of rotation. A high speed shaft is then connected to a generator that produces the electricity.

Solar photovoltaic systems

A solar photovoltaic system uses solar cells to convert light energy from the sun into electricity.

Energy ratings

Energy rating tables are used to measure the overall efficiency of a dwelling, with rating A being the most energy efficient and rating G the least energy efficient (see Fig 3.41).

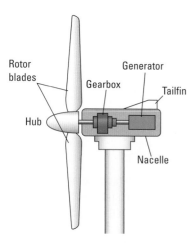

Figure 3.37 A basic horizontal axis wind turbine

Alongside this, there are environmental impact ratings (see Fig 3.40). This type of rating measures the dwelling's impact on the environment in terms of how much carbon dioxide it produces. Again, rating A is the highest, showing it has the least impact on the environment, and rating G is the lowest.

A Standard Assessment Procedure (SAP) is used to place the dwelling on the energy rating table. The ratings are used by local authorities and other groups to assess the energy efficiency of new and old housing and must be provided when houses are sold.

Preventing heat loss
Most old buildings are under-insulated and benefit from additional insulation, whether this is applied to ceilings, walls or floors.

The measurement of heat loss in a building is known as the U Value. It measures how well parts of the building transfer heat. Low U Values represent high levels of insulation. U Values are becoming more important as they form the basis of energy and carbon reduction standards.

By 2016 all new housing is expected to be Net Zero Carbon. This means that the building should not be contributing to climate change.

Many of the guidelines are now part of Building Regulations (Part L). They cover:

- insulation requirements
- openings, such as doors and windows
- solar heating and other heating
- ventilation and air-conditioning
- space heating controls
- lighting efficiency
- air tightness.

Building design
UK homes spend £2.4bn every year just on lighting. One of the ways of tackling this cost is to use energy saving lights, but also to maximise natural lighting. For the construction industry this means:

- increased window size
- orientating window angles to make the most of sunlight – south facing windows maximise sunlight in winter and limit overheating in the summer
- window design – with a variety of different types of opening to allow ventilation.

Solar tubes are another way of increasing light. These are small domes on the roof, which collect sunlight and then direct it through a tube (which is reflective). It is then directed through a diffuser in the ceiling to spread light into the room.

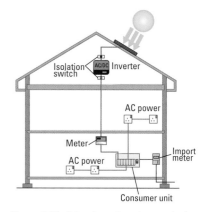

Figure 3.38 A basic solar photovoltaic system

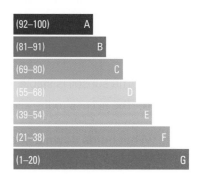

Figure 3.39 SAP energy efficiency rating table – the ranges in brackets show the percentage energy efficiency for each banding

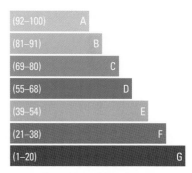

Figure 3.40 SAP environmental impact rating table

TEST YOURSELF

1. In which of the following types of building is a traditional strip foundation used?

   a. High rise

   b. Medium rise

   c. Low rise

   d. Industrial buildings

2. Which of the following is a reason for using a raft foundation?

   a. The subsoil is rock

   b. The subsoil is unstable

   c. The subsoil is stable

   d. The access to the site allows it

3. What holds down a floating floor?

   a. Nails and screws

   b. Adhesives

   c. Blocks

   d. Its own weight

4. What is another term for formwork?

   a. Shuttering

   b. Cavity

   c. Joist

   d. Boarding

5. What is the minimum distance the DPC should be above ground level?

   a. 50 mm

   b. 100 mm

   c. 150 mm

   d. 200 mm

6. A roof is said to be flat if it has a slope of less than how many degrees?

   a. 5

   b. 10

   c. 15

   d. 20

7. What shape is the upper part of a gable end?

   a. Rectangular

   b. Semi-circular

   c. Square

   d. Triangular

8. What do you call the horizontal timber that is placed at the top of a wall at eaves level in a roof, to hold the ends of joists or rafters?

   a. Fascia

   b. Bracings

   c. Wall plate

   d. Batten

9. What happens to the majority of construction demolition and excavation waste?

   a. It is buried on site

   b. It is burned

   c. It goes into landfill

   d. It is recycled

10. Which part of the Building Regulations 2010 requires construction to consider and use energy efficiently?

    a. Part B

    b. Part D

    c. Part K

    d. Part L

Chapter 4

Unit CSA–L20cc69

SET OUT MASONRY STRUCTURES

LEARNING OUTCOMES

LO1/2: Know how to and be able to interpret information for establishing setting out requirements

LO3/4: Know how to and be able to prepare for setting out masonry structures to the given specification

LO5/6: Know how to and be able to set out right-angled masonry structures on level ground to the given specification

INTRODUCTION

The aims of this chapter are to:

* help you interpret working drawings and other information

* help you to know how to set out regular shaped masonry structures on level ground.

INTERPRETING INFORMATION FOR ESTABLISHING SETTING OUT REQUIREMENTS

Interpreting and referring to different information in order to help you follow setting out requirements is a vital skill. If the setting out is incorrect then it will lead to mistakes that could prove very costly to put right.

There are a number of different types of working drawings, which we looked at in Chapter 2. You should refer back to the 'Drawings and plans' section of the book to refresh your memory.

Hazards associated with setting out masonry structures

Setting out needs to be carried out at an early stage of any construction job. It refers to the marking out and positioning of a building. It is a very important operation as it must be as accurate as possible. Mistakes made at the setting out stage would be very costly to put right later – for example, incorrect measurements could mean digging out finished concrete. Buildings have sometimes had to be completely demolished after being built in the wrong place!

This stage is known as 'front end construction work'. Before setting out, services will have been diverted and any demolition work will have been carried out. Setting out will happen before any **ground work** begins.

It is worth remembering that the area where the new structure is to be built may contain some unpleasant surprises underground. For example, rubbish may have been buried in the past. It is extremely important to make sure that you are careful when you clear the area ready for setting out.

In addition, there may be objects, such as cables, pipes, glass, metal and rubble, just under the ground that may cause problems when you put the first pegs in. A site inspection should identify any potential problems.

KEY TERMS

Ground work

– this is preparation work, such as clearing vegetation from the site, installing drainage and foundations. These are activities that must be undertaken before the rest of the construction can take place.

When you are preparing the ground, make sure that you wear appropriate PPE, such as gloves.

These are points of good practice:

* Refer to a method statement – this will have been written by the site manager and will explain precisely what needs to be done and how the job needs to be carried out. It will take into account any risks or hazards identified in the risk assessment.

* A risk assessment is another written document that identifies any possible risks or hazards in carrying out work such as setting out. It will explain exactly what tools, materials and equipment you will need, including PPE.

* Site safety rules – in addition to these two documents the site managers will be keen to ensure that all health and safety issues are considered. They may have ways of working that require even greater care than is legally required.

Drawings

When you are setting out masonry structures you will need to refer to different types of drawings. It is important to understand what all parts of the drawings mean. For example you will need to understand:

* abbreviations
* hatchings
* scales used.

Each type of drawing has a specific purpose and together they should provide you with the full picture of exactly what is required.

See Chapter 2 for more about interpreting drawings.

Purpose of the drawings

The purpose of drawings is to assist construction. They are organised in a logical sequence, which should follow the flow of the actual construction work. All details and sections are usually grouped together on the same sheet. Scales should be clear and show the details of the structure.

British Standards

BS 5964-3:1996 covers building setting out and measurement including setting out services that are needed at various stages of construction. The standard provides a number of checklists.

Other key British Standards include:

* BS 5964-1:1990 – methods of measuring, planning and organisation

* BS 5606:1990 – a guide to accuracy in building

* BS 7307:1990 – tolerances, methods and instruments used to measure and the positioning of measuring points

* BS 7334:1990 – optical levelling instruments and electronic distance measurement, including lasers and other optical instruments.

Discrepancies and inaccuracies

A bricklayer is unlikely to be responsible for either surveying or setting out the project. However, when they are a supervisor, the responsibility falls on them to check that the work follows what has been outlined in the drawings. It means knowing if something has gone wrong or could go wrong.

Once the setting out has been done it is always good practice to double check. Additional checks should be carried out before important new stages of construction get under way. If there are any differences, the drawing needs to be checked to ensure it is correct. This means talking to the architect or designer to clear up any misunderstandings and avoid potential problems.

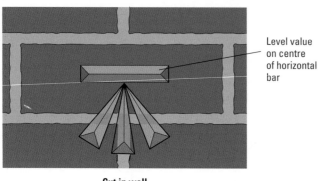

Level value on centre of horizontal bar

Cut in wall

Figure 4.1 Ordnance survey benchmark (OSBM)

Information sources

A wide range of different information sources can or should be used when setting out. Table 4.1 summarises these.

Datum	Explanation
Building Regulations (approved documents)	These are approved documents consisting of 14 technical parts of the Building Regulations. They are supported by good practice guides and set out minimum standards for the design, construction and alteration of buildings.
Local authority requirements (location of building line)	Local authorities have the power, dating back over 100 years, to set the building line. This is a conceptual line that extends along a street and indicates how close buildings can be constructed to roads/pavements/footpaths .
British Standards specifications and codes	British Standards have a range of specifications and codes of practice that must be followed during the setting out procedure.
Manufacturers' information	This can be found in manufacturers' catalogues or data and information sheets. It will suggest specific ways to lay out in order to achieve the maximum benefits from their products, such as blocks or damp-proof courses.
Ordnance Survey benchmarks (OSBM)	These are usually on public buildings and can be cut into walls. They are used as a marker to record the height against Ordnance datum and are an indication of the height above sea level.
Temporary benchmarks (TBM)	This is a known height from which any other level can be taken. TBMs are usually surrounded by a timber fence, concreted into the ground and often painted a bright colour.
Datums	In addition to site datum points, other datums can be useful as reference points. These can include foundations, damp-proof courses, oversite concrete and invert levels.
Site datum	This is a fixed point on the site that is usually located near the site office. This fixed point allows you to compare all other levels on site.

Table 4.1 Different sources of information

PREPARING CONSTRUCTION SITES

It is vital that construction sites are properly prepared before setting out activities can take place. Some elements of the site preparation, such as surveys, may have been arranged by the client or their architect, before the main contractor is involved. If there is a main contractor, they may be involved in the work at this stage.

The process usually begins with a survey of the construction site. This aims to:

* locate any existing utilities
* determine the boundaries
* carry out soil tests
* locate any underground obstacles.

During this process arrangements should also be made to:

* arrange for any fences or boards that may need to be erected
* demolish any derelict buildings and clear vegetation
* identify areas for site facilities and storage sheds.

This work is likely to be undertaken, or arranged, by the main contractor.

More detail on particular activities can be found in the next section.

PPE required for setting out

PPE should include:

* safety goggles or safety glasses – always necessary if there is a risk of injuring your eyes

* gloves – especially useful if you are working in cold conditions or where there is a risk of sharp objects sticking into your hands

* safety footwear – in case of sharp objects, uneven ground and poor weather conditions

* high visibility (hi-vis) clothing – so you can be seen easily on site

* hard hat.

Risk assessment

In Chapter 1 we examined the importance of identifying potential hazards and the ways in which risk assessments and method statements can help avoid possible health and safety hazards. It is also important to make sure that manufacturers' instructions are followed. All of these precautions ensure that you follow health and safety legislation and keep yourself safe.

KEY TERMS

Oversite concrete

– this is a layer of concrete, usually installed below a slab or other floor finish.

Invert levels

– this is the lowest part of any pipe, trench or tunnel that will carry liquids.

Figure 4.2 A walkover survey helps to identify potential problems

Clearance activities and walkover surveys

In the previous section we looked at some of the early stages that will become obvious in a simple walkover survey. This would help to establish the condition of the construction site.

The walkover survey will help identify any old foundations, drainage, subsidence and, importantly, any areas that might need further investigation. Each site will be different and some will need more clearance and preparation work than others. Table 4.2 outlines some common activities that may result from a walkover survey.

Site clearance issue	Explanation
Site planning	This can be very important because it identifies access, areas of the site where construction will take place and possible locations for site facilities and storage. It also identifies how the site can be made secure.
Positioning of resources	Building materials and equipment need to be placed and stored in a secure and convenient location. The walkover survey identifies possible locations. These need to have easy access for vehicles and still be in a convenient position so resources can be moved to the actual construction area.
Removal of obstacles	There can be a wide variety of different obstacles – vegetation, remains of old buildings or debris will need to be removed before construction can get underway. There may also be underground obstacles that need to be filled in and the site may need to be levelled.
Hedges and treetops	These could restrict access to vehicles and cause health and safety problems for those working on the construction site. Space is often at a premium, so hedges and low-lying branches may have to be cut back in order to provide the necessary space. Trees may have to be removed to make room on the site.
Flat and sloping sites	Flat sites may not present a particular problem in terms of access and siting of resources. Sloping sites can be more challenging, particularly if the construction is to be built into the slope. Temporary flat areas may have to be prepared and levelling of the site may be required.
Demolition and surface strip	If demolition is required then there may be limits as to when this can take place. In some cases, demolition vehicles may only be allowed to move in and out of the site on particular days. There may be restrictions on making excessive noise at certain times of day. There may be seasonal restrictions due to wildlife, such as bats, nesting in derelict buildings. There are also restrictions as to what can be done with the demolition waste. Surface stripping involves removing all surface vegetation, including the roots. This is usually up to a depth of 50 mm. Effectively it removes the root zone, as it is bad practice to build over surface vegetation.
Types of soil	According to Building Regulations there are seven different types of soil, plus subsoil. Local authorities can provide fact sheets to help work out foundations for different soil types found in the area. Topsoil is of particular importance because it is far less compacted than other layers of soil and usually has more organic matter, so is more unstable. It is therefore common practice to strip the topsoil off before construction. Some soil may be found to be contaminated, and additional treatment may be required.

Table 4.2 Site clearance issues

Locating and isolating existing services

The local authority should be able to provide you with information related to the location of services. However these services still need to be confirmed by walking the site. Sometimes it may be necessary to use surveying equipment to find services that are hidden. Existing services **must** be located and isolated before **any** work (including setting out) can be carried out.

Table 4.3 explains why it is important to locate and isolate existing services.

Existing services	Activity
Reinstatement	The provider of the service will usually be responsible for dealing with any new connections that are necessary. The main exception is drainage. Service providers will often demand that the ground work has been carried out to expose the services prior to reconnection.
Diversion of services	It will be necessary for certain services to continue to be available on the site and as a temporary measure they may have to be diverted along a different route. Perhaps the pipes and cables will be exposed for a time. These need to be protected and any trenches clearly marked.
Site safety	Exposed pipes and cables can be a problem and services should only be exposed for a limited period of time. Trenches should have warning signs and guard rails and heavy weights should not be placed close to the trench, as this could cause the trench to collapse. Some services, such as gas or electricity, could still be live and could cause a danger to those on site. Hitting live gas pipes or electric cabling could be highly dangerous. It may also impact upon buildings neighbouring the site, leaving them without services until they can be fixed.

Table 4.3 Dealing with existing services

There are a wide variety of different services, which could include:

* gas
* electricity
* water
* telephone
* electrical cable
* drainage (foul and storm).

Reclaiming materials

Construction is under pressure to become greener, and this means reclaiming as much demolition material as possible. Such materials can include:

* bricks
* slate roofing
* doors and window frames
* ceramic tiles
* metal fixtures and fittings
* glass panels
* cobbles
* timber
* steel.

Not only is reclamation becoming a legal requirement but it can also be a source of income: if buildings are demolished carefully the materials can then be sold on and reused via salvage companies. If you or your company want to sell on reclaimed materials, ensure that the client or customer has agreed to you selling materials from their site – often they will be grateful that they don't need to dispose of rubble themselves.

PRACTICAL TIP

Asbestos can be found in many older buildings, either as sheets, coating materials or insulation. Breaking up materials that contain asbestos can create asbestos dust. If inhaled it can be fatal. **Only licensed contractors are allowed to work with asbestos.**

Site clearance calculations

Most of the calculations simply require a little measurement and the use of a calculator. One of the most important jobs is to take the topsoil away from the **footprint** of the building. This might only take a couple of hours, but if there is sloping ground on the site it could mean taking away tonnes of top soil which might take days. This is the kind of simple calculation that can have a major impact on costs and timescales.

Figure 4.3 Digger excavator

Calculation	Description
Materials by volume and area	In order to estimate the volume of earth that is needed to be removed, think of the area as the footprint plus about 2 m of working space around it. Although the top 50 mm can be kept back to be reused as top soil, another 100 mm may need to be removed to accommodate the foundation. The cost depends on the land fill tax and is measured in m³.
Perimeter	In addition to the 2 m perimeter around the footprint of the building described above, any areas of hard standing or driveways further away from the building will also need to be cleared for future construction.
Quantities	A building with a 120 m² footprint would need to have around 30 m³ of excavated material removed. If there is a 5° slope on that same site then as much as another 200 m³ could be generated. Estimating volume and therefore quantity is very difficult, so thorough measurements need to be taken.
Costs	Costs will depend on the area and on the availability of lorry hire and other vehicles. A digger would usually cost around £6 per m³, as would a lorry, including the tipping fees. An additional £3 per m³ would be charged as land fill tax. This means that, depending on the area, and adding day rates for hiring machinery, removing 30 m³ is likely to cost up to £500 (2013 figures).
Mid-girth	The girth or thickness of trees and plants is measured at a height of 1 m from above ground level. The diameter of stumps is measured at ground level. For site clearances companies have a schedule of rates and prices depending on the girth.
Percentage for wastage and bulking	As already mentioned, wastage must be kept to a minimum, so working out precise figures can save a lot of money. On site, where waste is unavoidable, it is often not cost effective to arrange for its disposal until there is enough volume of material. It is possible to convert volume of construction waste into weight in kilograms. This is a more accurate way of working out potential costs.

Table 4.4 Calculations needed for site clearance

Basic surveying skills are needed and a close look at the drawings is important. Each wall will need foundations underneath it. At the very least timber, string and a level are required. A professional can usually complete the necessary measurements along with a colleague in a day. It is important that their measuring skills are accurate before any foundation trenches are excavated.

RESOURCES NEEDED FOR SETTING OUT WORK

There is a lot of work and checking to be carried out before construction can begin. You will need to set up profiles and take various measurements. You will also need to record readings so that you can refer to them later.

This important early stage of construction means that you can identify any potential problems and take action to put them right.

Checklist

It may not always be necessary to have all of these resources available to you while you are setting out. However, on many occasions much of this work *will* need to be done, and these various resources used, in order to successfully set out to build a masonry structure.

Resource	Description and use
Ordnance survey map 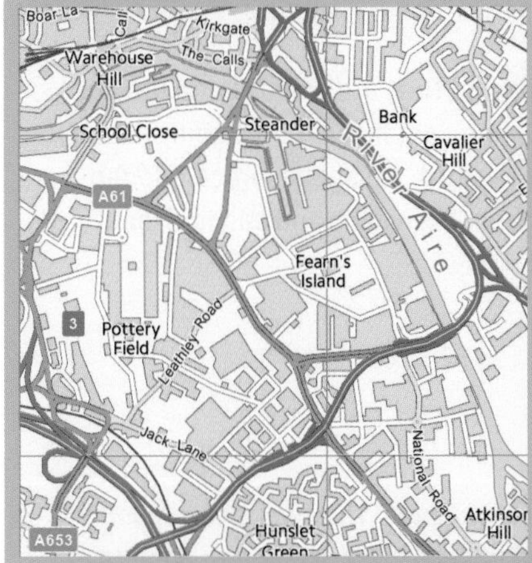 Figure 4.4	Ordnance Survey provides official surveying and topographic mapping.
Site plan	A site plan is a location drawing that shows the position of the building in relation to the site and its surroundings. See Chapter 2 for more details.
Block plan	These identify the site in relation to the surrounding area. See Chapter 3 for more details.

Resource	Description and use
Working drawings 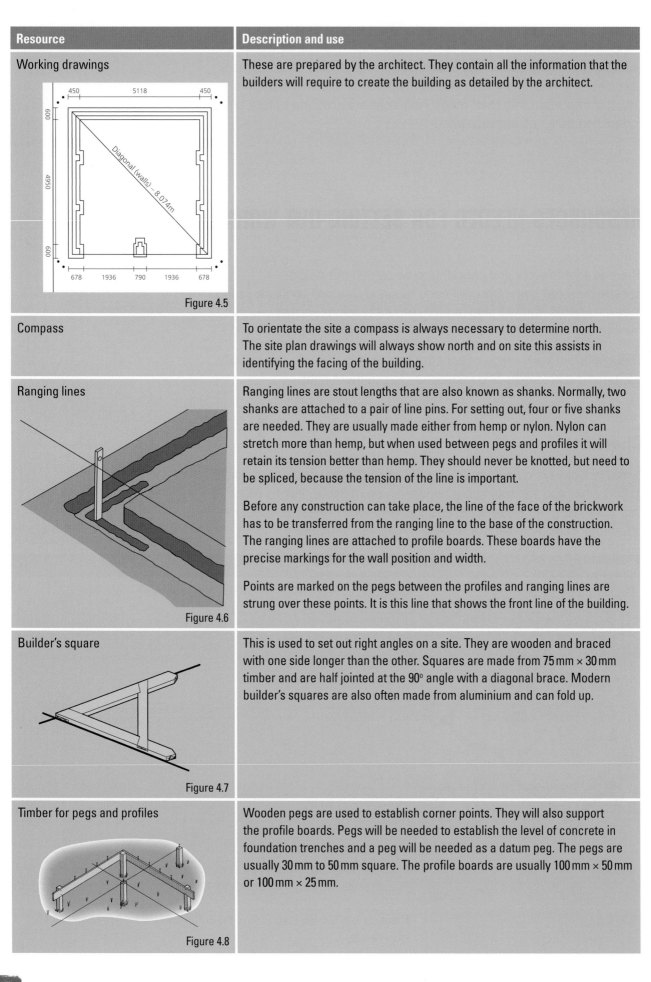 Figure 4.5	These are prepared by the architect. They contain all the information that the builders will require to create the building as detailed by the architect.
Compass	To orientate the site a compass is always necessary to determine north. The site plan drawings will always show north and on site this assists in identifying the facing of the building.
Ranging lines Figure 4.6	Ranging lines are stout lengths that are also known as shanks. Normally, two shanks are attached to a pair of line pins. For setting out, four or five shanks are needed. They are usually made either from hemp or nylon. Nylon can stretch more than hemp, but when used between pegs and profiles it will retain its tension better than hemp. They should never be knotted, but need to be spliced, because the tension of the line is important. Before any construction can take place, the line of the face of the brickwork has to be transferred from the ranging line to the base of the construction. The ranging lines are attached to profile boards. These boards have the precise markings for the wall position and width. Points are marked on the pegs between the profiles and ranging lines are strung over these points. It is this line that shows the front line of the building.
Builder's square Figure 4.7	This is used to set out right angles on a site. They are wooden and braced with one side longer than the other. Squares are made from 75 mm × 30 mm timber and are half jointed at the 90° angle with a diagonal brace. Modern builder's squares are also often made from aluminium and can fold up.
Timber for pegs and profiles Figure 4.8	Wooden pegs are used to establish corner points. They will also support the profile boards. Pegs will be needed to establish the level of concrete in foundation trenches and a peg will be needed as a datum peg. The pegs are usually 30 mm to 50 mm square. The profile boards are usually 100 mm × 50 mm or 100 mm × 25 mm.

Resource	Description and use
Measuring tapes	There are two different types of tape but they should both conform to British Standards. Steel tapes are generally the most accurate and are available in lengths of between 3 m and 100 m. The other type of tape is synthetic, which could stretch or shrink after use and exposure to the elements. They are usually either 20 m or 30 m in length.
Water level 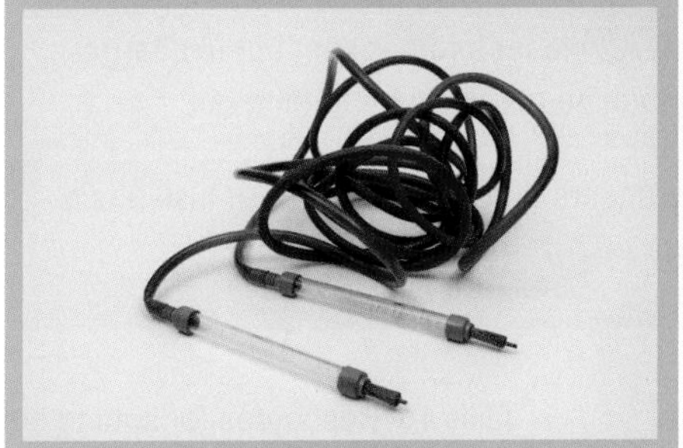 Figure 4.9	This is a length of hose that has a transparent tube in each end. The hose is filled with water. It is an ideal resource for checking levels in distances of over 30 m. It is a simple device, as water will always find its own level. A great advantage is that you can take levels around corners or obstructions. It is not normally used for setting out a site. It is normally used when transferring levels between rooms where walls are in the way.
Spirit level 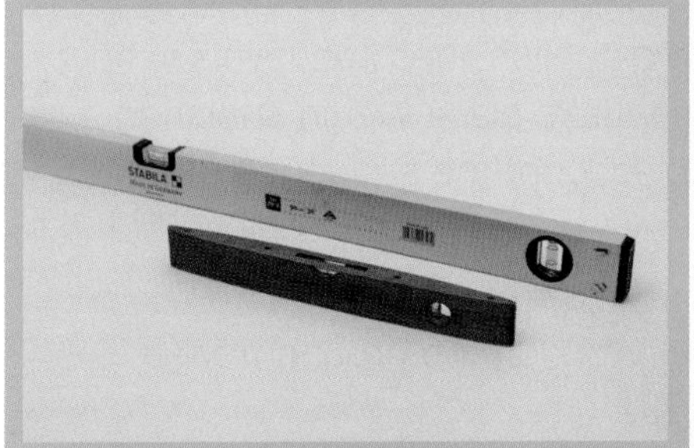 Figure 4.10	This is a simple straight edge that has a glass tube containing a liquid and a bubble of air. When the air bubble is in the centre of the tube the straight edge is exactly level. The tubes are marked with lines to confirm that the edge is level.
Straight edge 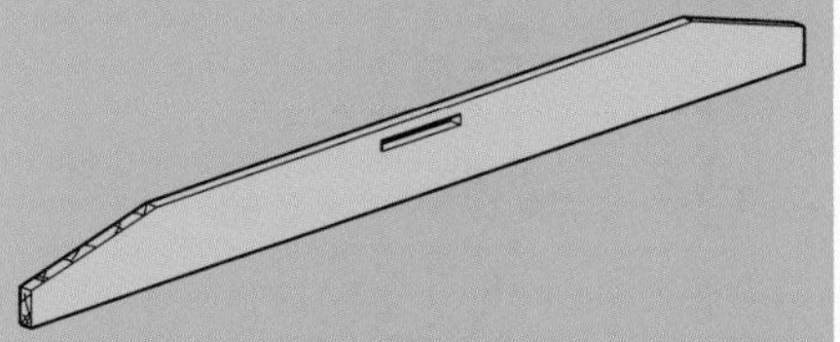 Figure 4.11	This simple piece of wood with parallel sides provides you with a straight edge, although not always an entirely accurate one.
Optical squaring equipment	A site square has two telescopes that are set at 90° angles to one another. This is mounted on a tripod. The square can be vertically adjusted and can provide you with a way of setting out right angles from 2 m to 90 m.

Resource	Description and use
Hand tools (hammers and saws) 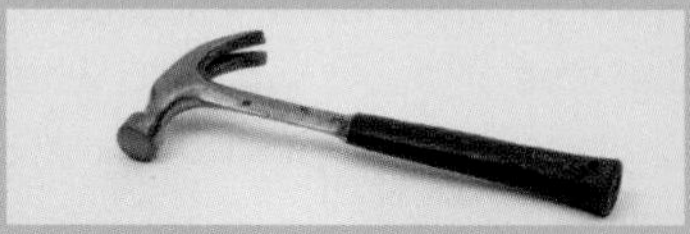 Figure 4.12	Primarily for preparing pegs and profiles, simple hand tools will be needed.
Optical laser level	A laser level works on almost exactly the same principle as an optical level but only needs one person to operate it. A laser beam is sent out to a target receiver. The level at that point can then be recorded for setting out.
Optical level	An optical level has a mechanism to ensure that it is horizontal. It also has a telescope which magnifies the image.

Table 4.5 Resources for setting out

KEY TERMS

Horizontal

– this allows you to set out the work from a drawing.

Vertical

– this is known as levelling and allows you to find the differences in elevation.

Selecting resources and making calculations

The key objective of setting out is to establish lines and levels in an accurate way, so that construction can take place within specified tolerances. It also needs to take into account the sequence of work.

There are two ways of doing this:

* On the **horizontal** – by setting out work on the ground from a drawing, as can be seen in Fig 4.13.

* On the **vertical** – by working out the differences in heights between two or more points, compared to a datum point (OBM) where you know the height, as can be seen in Fig 4.14.

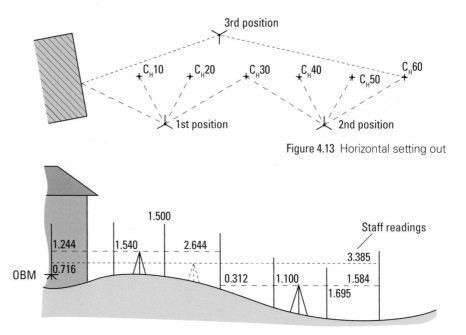

Figure 4.13 Horizontal setting out

Figure 4.14 Vertical setting out

Checking resources

All equipment and instruments are limited in their use and over time they may become less reliable. Relying on old, damaged or worn equipment and instruments could cause major problems when setting out. It is important always to test the equipment, as instruments in particular tend to be rarely checked. They should have an annual calibration test carried out by a specialist and certificated as accurate. You should always ensure that you set up instruments in the right way and, if you find inaccuracies, make allowances for them. Obviously, faulty instruments need to be repaired and damaged equipment must be replaced.

Protecting the work and surrounding area from damage

Refer to sustainability in Chapter 3 for more information on this.

SETTING OUT RIGHT-ANGLE SHAPED MASONRY STRUCTURES ON LEVEL GROUND

Setting out a masonry structure requires you to follow a series of key steps. These steps will ensure accuracy and correct positioning. For the setting out activity itself you will need to make sure that you have the correct equipment and materials:

* a steel tape measure, 30 m long with 1 mm graduations

* timber pegs and profile boards

* a ranging line

* a builder's square and a straight edge

* spirit and optical levels

* a site square

* hammer and nails.

Locating the position of the building line

Each local authority's planning department will have a fixed building line. This is the line beyond which a building must not go past. The position of the building line is a given distance from:

* the centre of the road

* the kerb line

* boundary walls

* existing buildings.

The building line can affect the building itself in a number of different ways:

* A building can be on the building line.

* A building can be angled toward the building line.

* A building can be some distance behind the building line.

* Some parts of the building may be allowed to go beyond the building line, such as a porch.

The building line can be found in the block plan. It is usually shown on site by pegs identifying the corners of the front of the building.

The procedure to establish a building line consists of the following steps:

1. A builder's square is placed on the rear edge of the kerb.

2. The position of the square is past the corner mark of the building.

3. The dimension is taken using a steel measure.

4. The dimension is transferred from the kerb to the building line.

5. The position is then marked with a wooden peg, usually with a nail showing the exact position.

Step 1

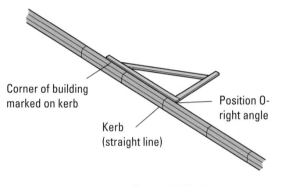

Corner of building marked on kerb

Position 0- right angle

Kerb (straight line)

Figure 4.15 Builder's square against kerb

Step 2

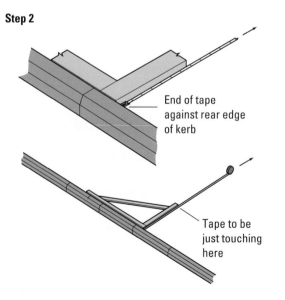

End of tape against rear edge of kerb

Tape to be just touching here

Figure 4.16 Transferring dimensions from the kerb to the building line

You will need to mark the line with a ranging line stretched between each of the pegs.

Locating the position of the site datum point

A site datum point is a fixed point on the site against which all other levels on the site can be measured. They can be used in order to carry out a variety of different activities, including setting the level of the damp-proof course.

Usually the site datum is related to another fixed point, such as:

* an Ordnance Survey Benchmark (OSBM) – see page 98

* kerb edge – you would normally knock a masonry nail into the joint behind the kerb so you are taking the reading from the same place every time

* frame cover.

The site datum is marked on the site by a steel post or peg concreted into the ground, usually near the site office.

Correct location

The first task is to determine precisely where the construction is located. We have already seen the value of block and site plans. But it is also important to ensure that you orientate the setting out of the building by using a compass to determine north. You can then refer to the plans, as these will have north marked on them.

Corner positions

It is important to identify the dimensions of the side of the building. This will enable you to set out the corner positions. There are several ways to do this. Fig 4.19 shows one way that this can be achieved.

The key stages are:

* run the ranging line from A beyond F

* using a builder's square, square the ranging line

* put in peg E and fix the ranging line to it

* measure from A to F

* put in peg F and knock a nail in it

* now fix the ranging line from pegs A to F.

Step 5

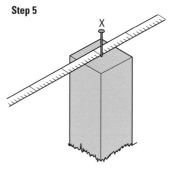

Figure 4.17 Peg with nail establishing correct position

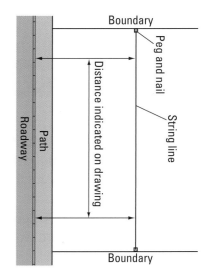

Figure 4.18 Building line and boundaries of site

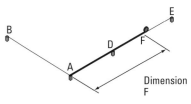

Figure 4.19 Setting out a right angle

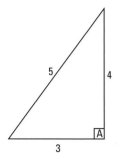

Figure 4.20 A triangle constructed with sides measuring 3, 4 and 5 units will produce a right angle at (A)

Right-angled corners

There are several methods of setting out a right-angled (90°) corner, some of which are shown in the practical tasks below. The 3:4:5 method of achieving right-angled corners uses a triangular shape to ensure that you produce a right angle. This can be seen in Fig 4.20.

The concept is straightforward. If you have a triangle that has sides with a ratio of 3:4:5 then a right angle will be created at A. The method is ideal for walls that are longer than 5 m.

PRACTICAL TASK

1. SET OUT A 90° CORNER USING A BUILDER'S SQUARE

OBJECTIVE

To make sure that all right-angled corners are square using a builder's square and checking it with the 3:4:5 method.

Ensure you select PPE appropriate to the job and site where you are working. Refer to the PPE section in Chapter 1.

TOOLS AND EQUIPMENT

Wooden pegs

Claw hammer

50 mm round nails

Two tape measures

Ranging line

STEP 1 Knock in two pegs (A and B) that are 1,500 mm apart. Knock nails into the centre of each peg, leaving enough of the nail sticking out so that you can attach the ranging line to them. Attach the ranging line. This indicates the face of the wall.

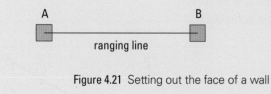

Figure 4.21 Setting out the face of a wall

STEP 2 Position the builder's square parallel with the ranging line to indicate the face of the wall.

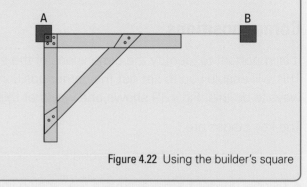

Figure 4.22 Using the builder's square

PRACTICAL TIP

The builder's square must be accurately lined up with the ranging line. One way of doing this is to support the square by laying down some blocks and resting the square on them.

STEP 3 Adjust the ranging line until it is parallel with the second leg of the builder's square.

STEP 4 Knock in peg C, knock a nail into the centre of the peg and attach the ranging line to it. Check again that the corner is square and make any minor adjustments if necessary.

PRACTICAL TIP

It is important at this stage to check the accuracy of the corner. One way of doing this is with the 3:4:5 method, by taking three straight lines (one 300 mm long, one 400 mm long and the third 500 mm long) and joining them together to make a triangle. The angle opposite the longest side will be 90°.

You can use any lengths of line, as long as you keep the proportion 3:4:5. The longer the line, the more accurate the corner will be, as long as you use the same unit to multiply each side. For example:

Multiply 300 mm × 3 = 900 mm
400 mm × 3 = 1,200 mm
500 mm × 3 = 1,500 mm

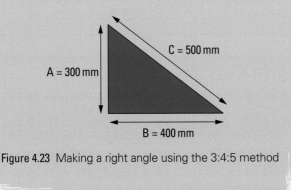

Figure 4.23 Making a right angle using the 3:4:5 method

STEP 5 Now measure 900 mm along the ranging line from the corner or nail in the peg A at the face of the wall and mark it on the ranging line. This can be done by tying a piece of ranging line tightly onto the face line. Then recheck the measurement.

STEP 6 Measure 1,200 mm along the second leg of the corner and mark it on the ranging line.as before.

STEP 7 Now measure between the two marks. The distance should be 1,500 mm if the corner is 90°. If it does not measure 1,500 mm then you will need to adjust peg C because you cannot adjust the face side.

You can also set out a corner using this method instead of a builder's square.

Optical site square
This contains two telescopes set permanently at 90° to each other. They can be adjusted vertically to enable you to look down at pegs that have been driven into the ground.

The site square is fixed to a tripod.

2. SET OUT A 90° CORNER USING AN OPTICAL SITE SQUARE

OBJECTIVE

To make sure that all right-angle corners are square using an optical site square.

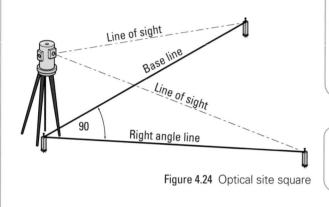

Figure 4.24 Optical site square

TOOLS AND EQUIPMENT

Three 50 mm × 50 mm timber pegs

Lump hammer

Optical site square and tripod

50 mm round nails

Ensure you select PPE appropriate to the job and site where you are working. Refer to the PPE section in Chapter 1.

STEP 1 Secure the site square to the tripod by using a large screw on the bottom of the site square.

STEP 2 Knock a nail into the top of the main corner peg and set the site square up over it. Site squares come with a steel rod or plumb bob (a line with a weight on it so that it will be plumb when hanging down). These are used to set the site square perfectly above the nail on the peg.

STEP 3 Using the bottom telescope of the site square, view along the building line to the other main corner peg. Adjust the site square until you can see the nail on the other peg in the centre of the cross hairs of the telescope.

STEP 4 Without moving the site square, look through the top telescope. This will give you a line at 90° to the building line. You can now knock in a peg into the ground in line with the view from the telescope and accurately knock a nail into the top of the peg in line with the cross-hairs on the telescope.

Dimensional accuracy

When you are taking dimensions from a drawing always ensure that you check whether they are accurate. If something is unclear then report it to the site supervisor. There should be an accuracy of 1:1000, which means the measurements should not be more than 1 mm out for every 1 m.

The drawings will have been produced to a British Standard and will have been approved by the local authority, so they should be correct.

Accuracy in setting out ensures that the correct dimensions are followed through to the finished construction.

Architects will normally have a disclaimer on the drawings which states that all dimensions must be checked on site and any discrepancies reported.

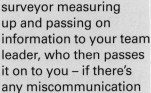

REED
TIP
!!!

Good communication is vital to getting a job done correctly. Imagine a surveyor measuring up and passing on information to your team leader, who then passes it on to you – if there's any miscommunication along the line, you'll end up building the wrong thing.

Transferring levels

We know that levelling is achieved by transferring levels from a point of reference known as a datum. The term levelling means transferring those levels to different positions on the site.

Levelling is all about establishing the relative heights of different points above or below the datum. The level is transferred from the nearest ordnance benchmark with a spirit level and straight edge. This will cancel out any inaccuracy in the level.

Using a spirit level, straight edge and wooden pegs, you start from a temporary benchmark and set up the straight edge towards the datum point. A temporary peg is driven into the ground and the level is used to check accuracy. The spirit level needs to stay in the same place on the straight edge. You simply continue the process until the datum peg position has been reached.

Temporary benchmarks (pegs) are the known height points and all other levels are taken from them. On site they should be protected, as all site levels are taken from this temporary benchmark (TBM).

A datum point can be established in relation to the temporary benchmark. It is an important reference point. The site datum is usually located at damp-proof course level. It will be marked by either a peg or steel post, which is concreted into the ground.

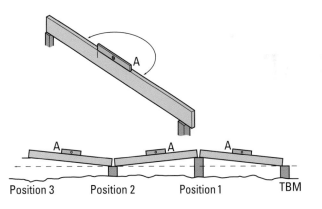

Position 3   Position 2   Position 1   TBM

Figure 4.25 Marking straight edge for spirit level position

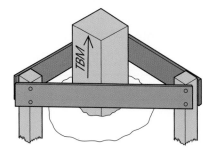

Figure 4.26 TBM and protective fence

PRACTICAL TASK

3. TRANSFER LEVELS USING A STRAIGHT EDGE AND SPIRIT LEVEL

OBJECTIVE

To transfer a given level to another point using a straight edge and spirit level.

TOOLS AND EQUIPMENT

Wooden pegs   Spirit level

Straight edge   Lump hammer

Ensure you select PPE appropriate to the job and site where you are working. Refer to the PPE section in Chapter 1.

All buildings need to be level so it is important to make sure that all the setting out is level at the beginning of the work. The simplest form of levelling is with a straight edge and spirit level, but this is normally only possible on smaller sites or over short distances.

This exercise requires points to be taken from a site datum. A site datum is a given position that all other datums on the site relate to. This can be items such as a kerb edge (with a nail inserted at the point), a window or door sill on an existing building, an inspection chamber cover or temporary wooden pegs knocked into the ground.

STEP 1 Knock a series of pegs into the area of ground that needs to be levelled. These pegs should be the same distance apart as the length of the straight edge. They will be level from the datum point or a set measurement above or below that datum point.

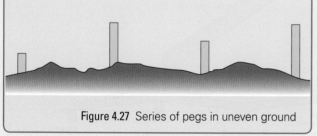

Figure 4.27 Series of pegs in uneven ground

STEP 3 Now reverse the straight edge and level (keeping the same edge on peg A), move the straight edge onto the next peg (B) and, using the level, level it with peg A.

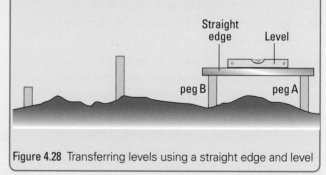

Figure 4.28 Transferring levels using a straight edge and level

STEP 2 From a given site datum, firmly knock the first peg (peg A) into the ground until it is approximately level with the site datum. Place the straight edge onto peg A and adjust it using the level until it is level with the datum.

PRACTICAL TIP

By reversing the straight edge and level you are cancelling out any discrepancies. For example, if the straight edge was 5 mm lower at one end (peg B) then if you do not reverse the straight edge then peg C would be 10 mm higher than peg A and so on.

STEP 4 Continue to level the rest of the pegs, reversing the straight edge as before.

Figure 4.29 Reversing the straight edge (1)

Figure 4.30 Reversing the straight edge (2)

Figure 4.31 Reversing the straight edge (3)

Figure 4.32 Reversing the straight edge (4)

Optical levels

When levels have to be transferred over a long distance, using a straight edge and spirit level can be very time-consuming and rather inaccurate. A more accurate method is to use a quick set level but you will need another person to hold the staff while you use the quick set level.

4. TRANSFER LEVELS USING A QUICK SET LEVEL

OBJECTIVE

To transfer a given level to another point using a quick set level.

Ensure you select PPE appropriate to the job and site where you are working. Refer to the PPE section in Chapter 1.

TOOLS AND EQUIPMENT

Wooden pegs	Staff
Quick set level	Lump hammer

STEP 1 Securely position the tripod and attach the quickset level to it using the screw fitted to the top of the tripod. The tripod has spikes in the three legs and these need to be stood on and forced into the ground to stabilise it and keep it in a secure position.

Figure 4.33 Quick set level

STEP 2 Adjust the instrument using the screws at the base until the fish eye is accurately positioned. The fish eye is a level set into the base of the instrument. You position the telescope section over two of the thumb screws and adjust them until the fish eye is in the centre. Now turn the telescope through 90° and adjust the thumb screw until it is level that way as well.

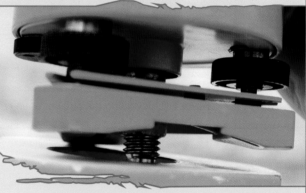

Figure 4.34 Setting up a quick set level – screw thread of fish eye

Figure 4.35 Quick set level showing the fish eye

PRACTICAL TIP

It is important that the tripod does not move once the quickset level is placed onto it. If it moves, the instrument will not be accurate. If you set the tripod up on hard ground then put something against the legs of the tripod to stop it moving.

PRACTICAL TIP

If you have to move, don't carry the tripod with the quick set level on it.

PRACTICAL TIP

When reading from a levelling staff, you start at the bottom of the letter E. In Fig 4.38 you can see 2.9 m on the staff and each section of the letter E is 10 mm. Therefore the top of the E is 2.95 m. Again, each square above the E is 10 mm. If the cross hair is between two sections then you have to estimate the measurement. For example on the staff the reading between 2.92 and 2.93 is 2.923 m.

STEP 3 Get a second person to put the staff onto the datum peg (see page 115) and aim the instrument at it. Adjust the focus until you can read the staff.

Figure 4.36 Using a quick set level and staff

STEP 5 Get the person helping you to move the staff to the next position to which the level is to be transferred.

STEP 6 The second person moves the staff up or down until the same reading appears in the cross-hair of the quick set level as on the previous position. This is marked onto the peg or the peg is knocked down until it is level.

STEP 4 Adjust the cross hair on the view finder so that you can take a reading from the staff and write it down.

Figure 4.37 Taking measurements on a section of staff

STEP 7 A timber cross rail can be nailed onto the peg so that the top of the cross rail will be in line with the transferred mark. This lets you rest the staff on it rather than trying to hold it against a pencil line, which could be rubbed or washed off if it rains.

STEP 8 Continue in this manner until you have transferred the datum level to all pegs.

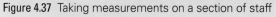
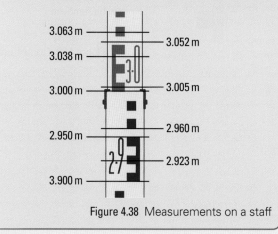

3.063 m
3.052 m
3.038 m
3.000 m
3.005 m
2.950 m
2.960 m
2.923 m
3.900 m

Figure 4.38 Measurements on a staff

PRACTICAL TIP

If you have to move the instrument, then you need to take a new reading from the last peg you levelled. If you are used to working alone then it would be better to use a laser level, which is set up just like a quick set level. Once the level is set up, it shoots out a red dot which you can position on the staff yourself.

Single and corner profiles

To keep ranging lines secure, profiles are used. These are kept in place by the line tension.

Profile boards are erected at each corner of the structure's footprint once the setting out is complete. They are used so that the corner setting out pegs can be removed to allow excavation work to take place. They are made from 50 mm × 50 mm timber pegs and 25 mm × 100 mm timber for profiles.

Normally the profiles are set at a datum level. This is usually the finished floor level (FFL) or damp-proof course level (DPCL).

PRACTICAL TASK

5. ERECT CORNER PROFILES

OBJECTIVE

To erect corner profiles using a profile board.

TOOLS AND EQUIPMENT

50 mm × 50 mm timber pegs

25 mm × 100 mm timber for use as profile boards

Lump hammer

Hand saw

Claw hammer

Straight edge

Spirit level

50 mm round head nails or screws

Ensure you select PPE appropriate to the job and site where you are working. Refer to the PPE section in Chapter 1.

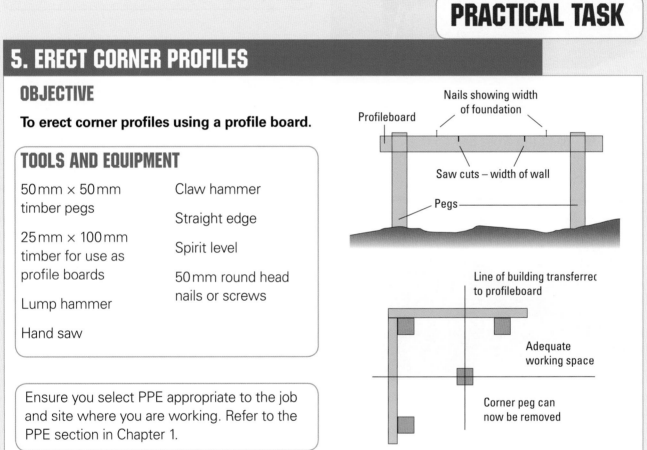

Figure 4.39 Corner profiles

STEP 1 Knock in two pegs at an appropriate distance away from the corner peg, parallel with the building line.

The distance away from the corner peg depends on the method of excavation. If you are excavating by machine you will need more room for the machine to manoeuvre, so approximately 1.5 m away from the corner peg. If you are excavating by hand then you can put the pegs a bit closer.

The distance between the two pegs should be the same as the length of the profile boards.

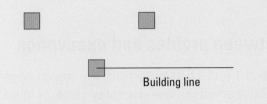

Building line

Figure 4.40 Two pegs running parallel with building line

STEP 2 Using one of the levelling methods previously completed, transfer the datum level to one of the pegs.

STEP 3 Using 50 mm round head nails or screws, attach the profile board to the pegs at the level marked on the peg. Make sure that the profile board is level between each peg.

STEP 4 Using a builder's square, knock in another peg at 90° to the other pegs. Refer to the practical exercise on page 112 if you need to refresh your memory on how to do this.

STEP 5 Using 50 mm round head nails or screws, attach the profile board to the pegs at the same level as the profile boards attached to the other peg. Make sure that the profile board is level between each peg.

STEP 6 Extend the ranging line fixed to the corner setting out peg and mark this line onto the profile board.

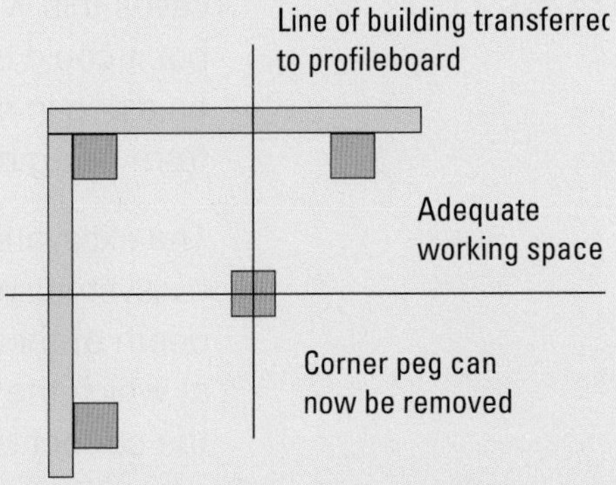

Line of building transferred to profileboard

Adequate working space

Corner peg can now be removed

Figure 4.41 Transferring ranging lines to the profile board

PRACTICAL TIP

It is not practical merely to leave a pencil mark on the profile board as this could be worn away. At this stage you can use either a saw cut or a nail to mark the face of the wall onto the profile. If you use saw cuts for the face of the wall, then use a nail for the edge of the foundation. Doing it this way, you will not get mixed up and think that the face of the wall is the foundation and vice versa.

STEP 7 You have now set out one corner. If you were setting out a building, you would re-check everything and complete the process for the remaining three corners. Once you are happy that all the dimensions are correct, you can remove the corner setting out pegs.

Walling and trench positions

Excavating foundations can begin once the area has been scanned for services and the ranging lines are in place, which indicate the width of the concrete. The first thing to do is to establish a datum level. In many cases this will probably be the damp-proof course on another building, but it could be from another datum point. The datum point is marked on a peg in the ground, which as we know is concreted in to prevent it from being moved.

The excavation of the trench can now begin. Once the trench has been excavated, you can knock steel pins into the foundation to indicate the depth of the concrete. The foundation concrete will come to the point at which the pins are fixed to the ground. This will ensure that you have the correct level.

Working space between profiles and excavation

Profiles should be between 1 and 1.5 m away from the trench edge. The actual trenches can be marked out using a spray paint, or in some cases, sand or dry cement.

There are several reasons to have working space between profiles and excavation:

* If machines are digging the trenches then there must be room for the bucket arm.

* If they are being dug by hand there must be space for a wheelbarrow.

* If the profiles are moved then once the concrete is in the walls may not fit the walling. This may be complicated and expensive to put right.

Datum heights

The datum height is important, as this has to be transferred in order to establish the foundation and wall positions.

Protecting setting out work

As we saw when we considered the working space requirements between the profile and the excavation, it is important that the profiles are not moved. They can easily become damaged and machinery could run over them. This would mean that all the measurements would have to be taken again. If damage or movement has not been noticed then there is a chance that the concrete will be put into the wrong position and the wall may not fit onto the foundations.

Accuracy, specifications, standards and problem solving

Inaccuracies can arise as a result of:

* incorrectly transferring lines and levels

* use of the wrong lines and levels

* calculations.

Accurate reference marks will enable you to mark the correct positions and the alignment and level of work. This can be achieved by using lines, pegs and nails.

British Standards lay out the basic building setting out and measurements and also suggest the tolerances for dimensional accuracy. It is important to bear in mind that bricks and blocks may not be 100 per cent dimensionally accurate (for example, they might be slightly shorter, longer or thicker than specified and may not have straight edges). If there are inaccuracies in the setting out this will only make the situation worse.

DID YOU KNOW?

British Standards specify the tolerances in the size of bricks and blocks. Making bricks is not an exact science. As they dry out they can shrink or be slightly out of shape.

CASE STUDY

South Tyneside Homes

South Tyneside Council's Housing Company

Get it right first time

Gary Kirsop, Head of Property Services at South Tyneside Homes, started off his career in the construction industry as an apprentice bricklayer.

'This is where your maths skills really come into play, because it's all about measurement and dimensions – taking them from a set drawing and then transporting them onto a live area where you're going to work.

You've got to make sure you're using the right scale rule (if you're using one) to take your dimensions off the drawing. But double check with the supervisor or site manager, if you're on a live site, because you want to be certain that what you start building is what's going to finish at the top of the structure, when the topping or roof goes on. So make sure that when you're taking your dimensions off the drawing with scale rule and putting it out on site, you're making your levels right as per the drawings – again, make sure it's correct.

In the real world there can be consequences for getting this wrong: the worst-case scenario is that you have to pull it all down, and you lose both time and money.

At the start of a project you get given a programme of works, and so most projects are built in a timely manner, and what you don't want is to find yourself in the worst-case situation. But even if you haven't had to pull it down, if the structure is too big, then there would be extra costs in heating the space.

It all comes back to making sure you take your time at the start, because that is what the client has asked for. They've commissioned an architect to do the drawings, and they want it that way, not your way. You could also be bringing long-term inherent problems into the structure, for example, getting cracks where you should have had an expansion joint. While it might not seem catastrophically wrong, there's still a risk of delayed building defects from any errors at this early stage.'

6. SET OUT A RECTANGULAR BUILDING

OBJECTIVE

To set out a rectangular building according to specification.

You need to establish the building line (see page 107). Great care should be taken on reading the drawings and making sure that the correct measurements are used.

The building line is 7 m from the kerb.

The distance from the boundary wall is 5 m.

The length of the building is 25 m.

The width of the building is 10 m.

> Ensure you select PPE appropriate to the job and site where you are working. Refer to the PPE section in Chapter 1.

TOOLS AND EQUIPMENT

The exact materials you will need to set out a building will vary depending on the size of the job. The following list would be adequate for, say, a small house:

50 mm × 50 mm timber pegs

25 mm × 100 mm timber for profiles

Lump hammer	50 mm round nails or screws
Hand saw	Cordless drill/screwdriver
Claw hammer	Site square (optional)
Builder's square	Ranging lines
Two measuring tapes (50 m), preferably steel	
Quickset level (optional)	

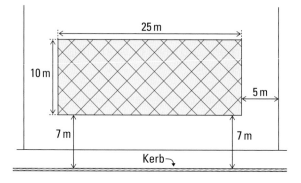

Figure 4.42 Plan of a house to be set out

STEP 1 Measure 7 m from the kerb and knock in two pegs (A and B), one at each side of the site (the plot of land on which the building will be erected). Knock a nail into the centre of each peg at exactly 7 m from the kerb. Use the nails to attach a taut ranging line between the two pegs.

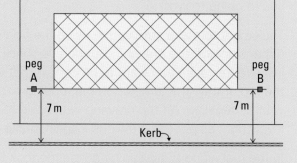

Figure 4.43 Establishing the building line

STEP 2 Place the main setting out corner peg (C) on the building line by measuring the given distance (5 m) from the boundary wall. Knock a nail into the peg at exactly 5 m.

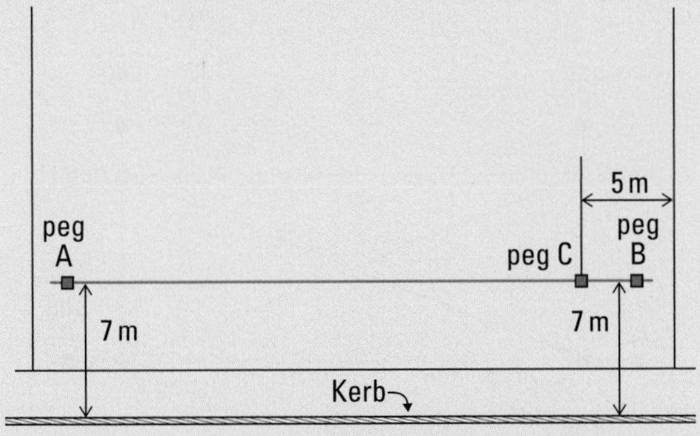

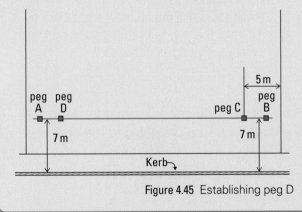

Figure 4.44 Establishing peg C

STEP 3 From peg C, measure the length of the building (25 m) and knock in peg D and putting a nail in it at the correct length of the building.

Figure 4.45 Establishing peg D

PRACTICAL TIP

It is good practice only to square one corner and set out the other walls using parallel measurements. This makes it easier to make any adjustments if the building is not square.

STEP 4 The width of the building is 10 m. From peg C use a builder's square to position peg E, greater than 10 m, at 90°. Attach the ranging line from peg C to peg E.

Using the 3:4:5 method (see pages 110–111), check the corner is square.

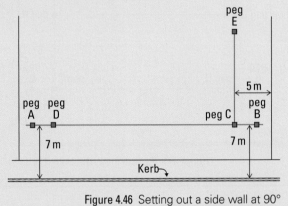

Figure 4.46 Setting out a side wall at 90°

STEP 5 From peg D, measure the same distance as the distance between pegs C and E and knock in peg F at 90° to peg D. Attach the ranging line from peg D to peg F.

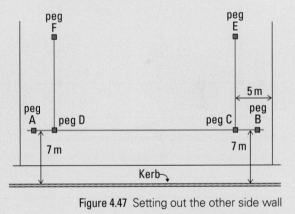

Figure 4.47 Setting out the other side wall

STEP 6 Now measure 10m from peg C in the direction of peg E and knock in peg H. Knock a nail into peg H exactly 10m from peg C. Repeat for pegs D to establish peg G.

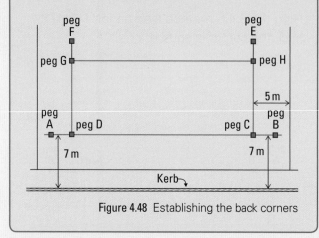

Figure 4.48 Establishing the back corners

STEP 7 Remove the ranging line from all the pegs and reattach it to pegs C, D, G and H. This gives you the outline of the building.

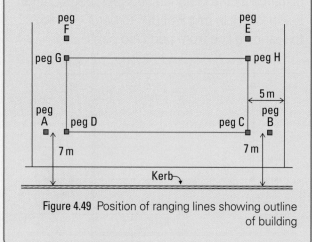

Figure 4.49 Position of ranging lines showing outline of building

STEP 8 You now need to check the diagonals to see if the building is square. Measure from pegs C to G and from D to H. The measurements should be the same if the building is square.

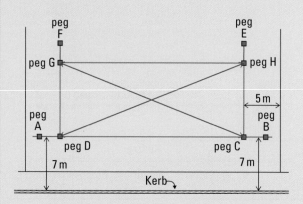

Figure 4.50 Checking the diagonals to see if the building is square

PRACTICAL TIP

If the building is not square DO NOT MOVE THE RANGING LINE but adjust the back line by moving pegs G and H the same distance on the peg or move the pegs if necessary.

If you have made any adjustments re-check all the dimensions before going to the next step.

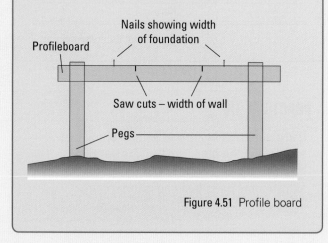

Figure 4.51 Profile board

PRACTICAL TIP

Now that the building is set out correctly, someone with a machine is going to dig the foundations. The pegs you have set out are in the way so if you don't move them, they will be dug up and all your work will be wasted.

You must move the pegs using corner profiles, which consist of two pegs and a profile board on which you mark out the position of the wall and foundations (see the practical exercise on page 110). They must be positioned well away from the corners; the distance depends on the method of excavation. If the foundations will be dug by hand the profiles will be approximately 1 m but if they will be dug with a machine then the profiles need to be at least 1.5 m away to give the machine plenty of room to manoeuvre.

STEP 9 You have to extend the lines that are attached to the pegs so that they pass over the profile boards. The position can be now marked by either using a nail or putting a saw cut into the profile board.

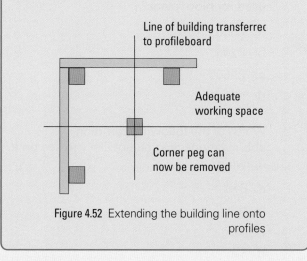

Figure 4.52 Extending the building line onto profiles

STEP 10 Once you have attached the line to all the profiles check the measurements, including the diagonals, to see if the building is still square.

TEST YOURSELF

1. Which of the following scales are usually used for block plans?

   a. 1:500

   b. 1:200

   c. 1:2500

   d. 1:5

2. Which of the following common abbreviations are you not likely to see on a drawing?

   a. Bwk

   b. UPVC

   c. Rwg

   d. Svp

3. Which organisation would usually be involved in determining the building line?

   a. Local authority

   b. Ordnance Survey

   c. Trading Standards

   d. British Standards

4. Who would normally be responsible for reconnecting services on a construction site?

   a. The main contractor

   b. The architect

   c. Local authority

   d. The service provider

5. At what height is the girth or thickness of trees or plants usually measured from ground level?

   a. 0.5m

   b. 1m

   c. 1.5m

   d. 2m

6. What is the term used to describe a fixed point on a construction site against which all other levels can be measured?

   a. Site datum point

   b. Site peg

   c. Site post

   d. Site midway position

7. Which shape is most closely associated with the 3:4:5 measurement of right-angled corners?

   a. A square

   b. An oval

   c. A triangle

   d. A circle

8. Ideally, to what other use could a spirit level be put?

   a. Skimming plaster

   b. Straight edge

   c. Pegging out

   d. Measuring distances

9. When establishing a building line, which piece of equipment is first used and placed on the rear edge of the kerb?

   a. Hammer

   b. Spirit level

   c. Builder's square

   d. Tape measure

10. What piece of equipment is useful for transferring levels over a long distance?

    a. Steel measuring tape

    b. Long length of timber

    c. Straight edge and spirit level

    d. Optical level

Chapter 5

Unit CSA–L2Occ72

CONSTRUCT SOLID WALLING INCORPORATING PIERS AND ARCHES

LEARNING OUTCOMES

LO1/2: Know how to and be able to prepare for constructing solid walling

LO3: Know how to set out to construct solid walling to the given specification

LO4/5: Know how to and be able to construct solid walling to the given specification

LO6/7: Know how to and be able to construct isolated and attached piers and arches to the given specification

INTRODUCTION

The aims of this chapter are to:

* help you to select materials, components, tools and equipment

* help you to build solid walls incorporating isolated and attached piers and arches.

PREPARATION FOR BUILDING SOLID WALLING

Traditional solid walling was the most popular construction method for buildings until the 1920s. The walls had a single solid wall around 500 mm thick. The walls would:

* support the roof

* support the upper floors

* keep the building warm

* prevent water from getting into the building.

The problem was that the solid walls were capable of supporting the roof and floors, but because of poor workmanship they often failed on the other two purposes. Even larger and more expensive buildings had major problems with water penetration. The walls became damp and timbers rotted on the floors and roof spaces. The lack of insulation meant that in poor weather the buildings were very cold and draughty. As a result, cavity walls became more popular and in time insulation was added to the cavity.

However, there are benefits to solid walling. The depth of wall offers its own insulation and it keeps the property cool in warm weather. The fact that the load is spread makes the whole structure stronger, so foundations are less likely to fail.

The drawback, however, is that solid wall construction is expensive. Additional, costly problems still need to be addressed, including the foundations and the damp-proof course. As a result it is not common to construct buildings using solid walling.

Solid wall preparation

Refresh your memory on the following:

* potential hazards

* types of drawings and how to interpret them

* risk assessments

* PPE

* protecting materials, completed work and the environment

* preparing and cutting components.

There are other areas that require specific information.

Drawings and conventions

Drawings are covered in detail in Chapter 2.

The architect may use **computer-aided design software**, or may draw the diagrams themselves. Sketches are simply that – they describe the particular requirements of either a construction procedure or a component.

KEY TERMS

Computer-aided design software

– this is also known as CAD or computer-aided drafting. This software allows the user to create a technical drawing.

Specifications and schedules

You should always check drawings, specifications and schedules and ensure that you use measurements from the site drawings.

Specifications are closely linked to working drawings. They provide additional information that is not usually on the working drawings. They could highlight any particular problems; they might detail work required and cover other issues, such as working hours on site or whether the site has limited access.

Other features that can be found on the specification form may relate to the quality of materials that are to be used.

A schedule can be either prepared by the builders themselves or by the designers on a larger job. The schedule is also used alongside working drawings. Schedules are useful for the following reasons:

* They detail the quantities of any materials that will be needed.

* They detail the physical characteristics of materials, such as type or size.

* They indicate where the materials will be placed on the site.

An example of a schedule can be seen in Fig 5.1.

DOOR SCHEDULE – INTERNAL ORDER NO.2431 2601			CONTRACT COMMON FARM ESTATE								

Supplier: William Nolan Ltd.

Item	Total Qty	Type	Size	Hand	Lock Latch	Hinge	Wall	Fanlight	Threshold	Head Height	Remarks
1	70	D7	2040 × 526	LH	1521 Roller catch	2 No. L/Pin	Frame 57	6 mm plywood	NIL	2374	Adjustable head 10 mm clearance between bottom of door and bottom of frame
2	72	D7	2040 × 526	RH	1521 Roller catch	2 No. L/Pin	"	"	NIL	2374	"
3	163	D9	2040 × 726	LH	915 Tubular latch	2 No. L/Pin	"	BDS	NIL	2374	
4	162	D9	2040 × 726	RH	"	2 No. L/Pin	"	BDS	NIL	2374	"
5	72	D9	2040 × 726	LH	2294–3 Bath lock	2 No. L/Pin	"	BDS	NIL	2374	"
6	74	D9	2040 × 726	RH	"	2 No. L/Pin	"	BDS	NIL	2374	
7	75	D9	2040 × 726	LH	1521 Roller catch	2 No. L/Pin	"	6 mm Plywood	NIL	2374	
8	75	D9	2040 × 726	RH	"	2 No. L/Pin	"	"	NIL	2374	"
9	34	D10	2040 × 826	LH	915 Tubular latch	2 No. L/Pin	"	BDS	NIL	2374	"
10	40	D10	2040 × 826	RH	"	2 No. L/Pin	"	BDS	NIL	2374	"
11	16	D13	2040 × 726	LH	"	2 No. L/Pin	"	NIL	NIL	2082	10 mm clearance between bottom of door and bottom of jamb
12	17	D13	2040 × 726	RH	"	2 No. L/Pin	"	NIL	NIL	2082	

A typical door schedule

Figure 5.1 Schedule form

Discrepancies

Occasionally there may be mistakes on working drawings or other documents. These can affect the way in which you will have to carry out the work. Typically, the following problems could occur:

* The scale and therefore the measurements on the drawings may be inaccurate once you have begun to measure out and set out the work.

* Vital information may be missing. A document may not have been provided, or information may be missing from one of the drawings or documents.

* The information on one of the drawings may be different from another document, such as the specification.

In all cases it is sensible to refer back to the individual who prepared the working drawings or documents and ensure that the errors are not carried through to the construction work.

You need to refer to the drawing number and date to make sure you have the latest drawing and/or specification when making reference back to the originator to clarify any details.

Work method statements

To remind yourself about method statements, refer to Chapter 1.

Resources required

The specification that you are using will tell you what types of materials are to be used. They will also tell you the joint finish.

Using the plans you can work out the materials you will need for the job. You will also need a range of brick working tools and the correct PPE. Before you get underway with the work you need to check that your materials and your equipment are ready for use. You should always ask yourself the following questions:

* Do I have the right materials and are there enough of them?

* Are the materials in good condition or are they damaged?

* Do I have all the tools that I will need for the work and are they in a good state of repair?

* Do I have the necessary PPE and has it been maintained properly?

PRACTICAL TIP

The Health and Safety Executive (HSE) has information about different types of work method statements for different types of construction tasks. It is also possible to find and buy completed work method statements from HS Direct at www.hsdirect.co.uk.

Calculations and formulae to determine quantities of materials, components and fixings

In order to work out the amount of materials required for a particular walling job, you need to know the area of the wall. This is the surface that it covers. You can do this by multiplying the length by the width to give the area of the walling.

For some jobs it can be more complicated, as there may be separate walls of different shapes. In this situation you have to work out the area for each shape and then add them all together.

Working drawings will usually show lengths and widths in millimetres, so these need to be converted into square metres. It is often a good idea to convert from millimetres to metres before you start making any calculations.

If a working drawing shows 4,000 mm then you will need to divide it by 1,000. This is because there are 1,000 mm in each metre so 4,000 mm becomes 4 m.

Example

On a working drawing a wall is shown as being 9,500 mm long and 2,500 mm high. The area of the wall in metres is therefore 9.5 m × 2.5 m = 23.75 m.

The formula to use is L (length) × H (height) = A (area)

As a rule of thumb, it is best to assume that there are 60 bricks per square metre.

For a half-brick wall, once you know the total area of the wall that you will be constructing you multiply the number of square metres by 60.

Bricks are made in standard sizes so 60 bricks per m² is a good rule of thumb. However, it is not always as simple as this, as there are other issues, which include:

* taking into consideration openings, such as doors or windows

* checking that all the measurements in the different documents and drawings have all been converted to metres

* adding extra bricks to the quantity in case any of them are damaged

* adding in additional bricks if you have one brick walling or one and a half brick walling. The calculation of 60 bricks per square metre is based on half brick walling.

Example

A customer wants a half brick wall that is 6 m long and 3 m high. Assume that 5 per cent of the bricks will be damaged. So, 6 m (length) × 3 m (height) = 18 m²

18 × 60 (standard number of bricks per m²) = 1,080 bricks

> **DID YOU KNOW?**
>
> Half brick walling uses 60 bricks per square metre. One brick walling uses 120 bricks and one and a half brick walling 180 bricks.

> **DID YOU KNOW?**
>
> For ordinary bricks, builders will often factor in a waste of around 5 per cent. For facing bricks this is 6 per cent.

To calculate the 5% wastage, (5 ÷100) × 1,080 = 54 bricks

1,080 + 54 = **1,134 bricks required**

Example

A customer wants a one and a half brick wall that is 12 m long and 2.5 m high. So,

12 m (length) × 2.5 m (height) = 30 m²

30 × 180 (standard number of bricks per m²⁾ = 5,400 bricks

Wastage (5 ÷100) × 5,400 = 270 bricks

270 + 5,400 = **5,670 bricks required**

Resources

A number of key resources are needed in order to carry out the construction process. We will look briefly at each of these resources, although it is important to remember that not all of these resources may be needed for every construction job.

Mortars

Mortar hardens as it dries. It is a mix of sand and cement or plasticiser, to which a quantity of water is added. It needs to be sufficiently plastic in order to spread in a relatively thin layer. Mortar is made up as follows:

* Cement – when it is mixed with water it will harden and bind the aggregate mixture.

* Aggregate – this is inert and is usually sand, gravel or crushed stone.

Cements

A number of different types of cement can be used. Some of these are more or less suitable for thin joint masonry and masonry cladding and in particular for use in thin joint blockwork, as can be seen in Table. 5.1.

Type of cement	Features and suitability
High-alumina	Quick-hardening and strong. High cost, therefore not widely used. Must not be used for structural work or masonry.
Masonry	Highly suitable for blockwork and rendering. Good plasticising properties and very workable.
Ordinary Portland	Widely used for block laying and has many applications. Ideal when there is not a high risk of either thermal or chemical impacts on the construction.
Rapid-hardening Portland	Hardens more quickly than ordinary Portland cement and has higher lime content.
Sulphate resisting	Ideal for below-ground concrete where damp may be an issue.

Type of cement	Features and suitability
Super sulphate	Highly resistant to sulphates.
	Must not be mixed with other products or cracking can take place.
White Portland	Usually used for rendering and mortar.
	Pigments can be added to it.

Table 5.1 Different types of cement

Lime

Figure 5.2 Cement

Lime mortar is generally only used on heritage work with ratios to sand of around 1:3, depending on various factors. There are three different ways in which lime can be added to the mortar, each of which has its own characteristics:

* Slaked quick lime (calcium oxide) – water is added, making the quick lime hot. The quick lime becomes putty-like and saturated. It is then added to the sand to create a lime mortar. It is used in conjunction with cement for modern work.

* Hydrated lime – this is added in powdered form to the sand before any water. It improves the water retention of the mortar and its workability.

* Hydraulic lime – NHL2, NHL3.5 and NHL5 are used for different purposes, mostly in older properties.

DID YOU KNOW?

Marble, limestone and chalk are all types of calcium carbonate. These are burned to create calcium oxide or quick lime.

Sand

For use with mortar, sand needs to be clean, sharp and well-graded. In fact, the sand's properties will have a huge impact on the quality of the mortar. The sand always needs to be clean, sharp and graded:

* Clean – it should not have any impurities. Dirty sand will have an impact on the adhesive properties of the mortar. You can wash sand on site if necessary.

DID YOU KNOW?

With experience, you will be able to instantly know whether the sand is sharp enough by rubbing the grains between your hands.

* Sharp – sand from the sea or from rivers has rounded grains and affects the adhesive qualities of the mortar. Sea sand will also have salt in it which needs to be washed to remove the salt as the salt is a contaminant. Suitable sand comes from quarries or from pits.

* Graded – the grains need to be of different sizes. They should not be too fine or too coarse. The more varied they are, the better the texture of the sand and the finished mortar.

Figure 5.3 Building sand

Premixed/retarded mortars

The big advantage of using premixed mortar is that it is accurately measured, has the right proportions and is made under quality-controlled conditions. There are two basic choices:

* Wet – this arrives on site ready to use and can be stored on site.

* Dry – this requires the addition of water.

A retarder slows down the reaction between the cement and the water by reducing the rate of water penetration. This means that the mortar is workable for a longer period of time. On site, a retarder can be added to mortar, or it can be already in the mortar received from the supplier.

PRACTICAL TIP

Wet ready-mix is easy to store and protect in tubs. Dry mix is stored in bags or silos which need to be moisture-free.

Water quality

Impurities in water can have an impact on the quality and strength of the mortar. Potable water (drinking water) is usually the best choice. Any contamination in the water can affect the strength of mortar and the overall strength of the wall that is being built; it can also cause rusting in any metals used in the construction. The presence of any of the following can cause problems:

- salt
- alkaline
- organic materials
- vegetable growth
- acid
- sugar
- oil
- sewage.

Figure 5.4 Mortar at the correct consistency

Mortar additives

Several different types of additive can be mixed with the mortar to produce different results. Plasticisers are an air entraining agent which improves the workability of the mortar. Accelerators reduce the time it takes for the mortar to set by increasing the speed of the chemical reaction. Colouring agents are dyes or **admixtures**. They are passive pigments, which add colour to the mortar without changing anything else.

Plasticisers

There are three different types of plasticiser, which are outlined in Table 5.2.

Plasticiser type	Description and use
Mortar	This is known as a liquid air entraining mortar plasticiser. It reduces the amount of water required to make the mortar workable. As less water is used it means that the mortar is more resistant to hot and cold temperature changes.
Chloride-free frost proofer	This allows you to work at below zero temperatures on the construction. The material accelerates the setting time.
Powder	This helps to protect the mortar from frost and makes it more durable. It is often used as an alternative to lime.

Table 5.2 Different types of plasticiser

Accelerators

Accelerators allow you to work in lower temperatures. They also strengthen the mortar and can be used as a frost proofer.

> **PRACTICAL TIP**
>
> If you are in doubt about the water quality it should always be tested at a laboratory. The pH should not be lower than 6.

> **KEY TERMS**
>
> **Admixtures**
>
> – aim to change the properties of mortar. There are three types. Active ones react with a component of the mortar, e.g. a plasticiser. A surface active one affects the reaction with either the air, water or a solid, e.g. an accelerator. A passive one, which has no impact on the mortar, only changes its colour.

Colouring agents

Pigments are inert so they do not react with other components of the mortar mix. The best pigments are alkali-resistant. This means that their colour will not be affected if alkaline is present. They should be light-fast, so that the colour does not fade when they are exposed to ultraviolet radiation from the sun. They also should be environmentally stable, as the pigments will be exposed to different temperatures and humidity over time.

Adhesives

Adhesives are basically bonding agents. They have two main purposes:

* to improve the adhesion of renders

* to improve the adhesion of levelling screeds.

Usually they are made from polymers:

* styrene-butadiene polymer (SBR), which is often better known as latex

* polyvinyl acetate (PVA).

These are emulsions or milky liquids and under normal circumstances they replace half of the water used in the mortar mix. These bonding agents, or adhesives, need to be added in precise quantities, particularly if you are working below the damp-proof membrane.

Tools and equipment

All the tools that bricklayers use need to be:

* capable of withstanding a great deal of use and wear

* capable of giving many years of service

* used for the job for which they are intended.

You will need the following tools for walling, as described in Table 5.3.

PRACTICAL TIP

Cheap tools are not a good buy. It is worth investing in tools made by well-known manufacturers that have a good reputation. On many sites you will be judged by the quality of your tools and their appearance.

Tool	Description
Bricklayer's trowel and pointing trowel	Trowels are made in sizes from 225 to 350 mm. A 250 to 275 mm blade is ideal for general work. There are left and right-handed blades. Pointing trowels range from 75 to 150 mm.
Spirit level	These are available in a variety of different lengths of up to 1.8 m. Smaller levels are ideal for adjusting individual bricks when doing decorative work. Longer ones are good for plumbing courses.
Brick hammer	These can have a hammer at one end and a blade at the other. Some will have a slotted end into which either a blade or a comb can be inserted.
Jointing iron	These are usually 60 to 125 mm long and used to form recessed joints.
Brick bolster and lump hammer	Bolsters are used with lump hammers to cut bricks. Lump hammers are around 1 kg and usually heavy enough for cutting away holes in brickwork.
Line and pins	These are made of steel and bought in pairs. The line is bought in knots. For general purposes two knots of line are sufficient on one pair of pins.
Plumb bob	Plumb bobs are usually made of lead. They are used to measure the vertical straightness of brickwork. They are rarely used these days.
Brick tong	These are gadgets that allow you to carry several bricks, usually 6 to 10, at one time in a safer way than trying to balance them. You lift a clamp and jaws clamp around the bricks. They are usually metal and fully adjustable.

Table 5.3 Different tools required for walling

SETTING OUT TO BUILD SOLID WALLING

When you are constructing solid walling you will need to understand how bricks and blocks have to be arranged. Together they are tied or untied as a result of the type of bonding. There needs to be good bonding throughout the wall as this ensures that the wall is a solid piece and also that the load is distributed in the correct way.

Positioning and locating bricks, blocks and other components

It is good practice to have all of the materials that you need for immediate use close at hand. You should certainly try to have sufficient components to last you for a set period of work, such as a half or full day.

Establishing bonding arrangements

Bonding is the way in which bricks or blocks are arranged in order to tie them together. Bonding has three main purposes, i.e. to make sure that:

* the wall is strong and any loads on it are evenly distributed

* the wall has lateral stability and it is resistant to any side thrusts

* the wall is attractive and even.

Bricks need to be laid out so that they have maximum strength. They can be laid to half bond, which means that the brickwork overlaps half of the brick below it. The minimum overlapping should be a quarter bond (the brickwork overlaps a quarter of the brick below it).

If the brickwork is an independent wall then the second course of brickwork should begin with a half brick. With returns, the bricks should have a flush overlap and a small cut brick placed next to it, in order to ensure that there is a half bond. This is why the choice of brick or block-cutting equipment is very important. Efficient and timely production of cut bricks is necessary to ensure that building the wall is not slowed down.

Establishing face bonds

As we have seen, brickwork bonds have a number of purposes, such as to:

* give strength to the wall

* make sure that loads being carried are distributed over the wall

* give lateral stability

* make the wall resistant to side thrust

* give a neat and presentable appearance to the wall.

In order to understand bonding it is important to appreciate some basics about bricks. These can be seen in Table 5.4.

Term	Explanation
Work size and brick terms	The standard brick is 215 × 102.5 × 65 mm. In a wall, if the 102.5 × 65 mm face is exposed then the brick is called a **header** (the end of the brick). If the 215 × 65 mm face is exposed it is a **stretcher**. The surface underneath the brick is the **bed**. The depression at the top of the brick is a **frog**. The sharp edge is called the **arris**.

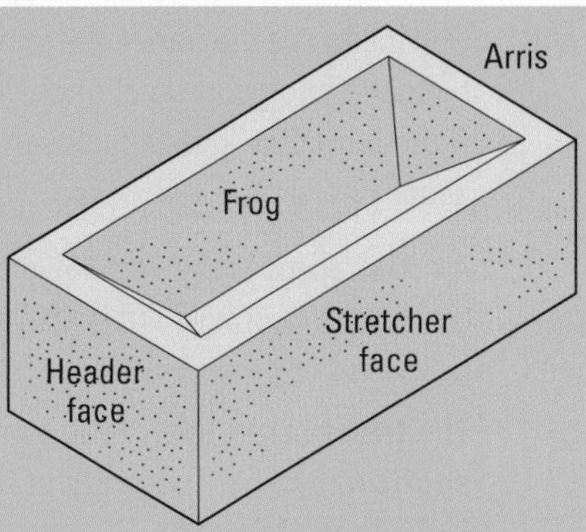

Figure 5.5 Brick with terminology

Coordinating size	Given the fact that mortar joints are added to each of the dimensions of the bricks, the actual coordinating size is larger than a standard brick size. Usually a 10 mm joint is added. This means that the coordinating size is 225 × 112.5 × 75 mm.

Sizes

	Length	Width	Height
Coordinating	225 mm	112.5 mm	75 mm
Work	215 mm	102.5 mm	65 mm

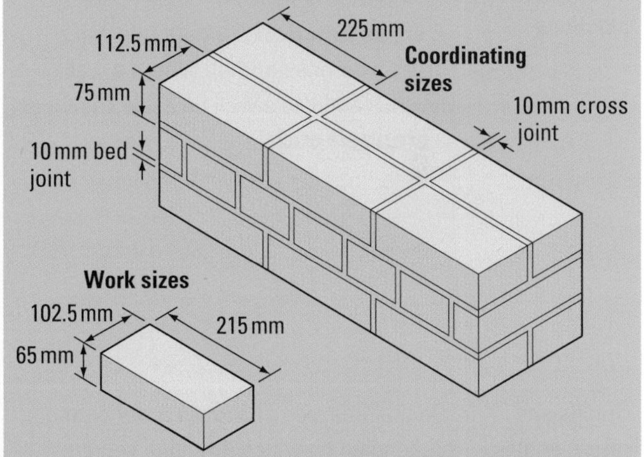

Figure 5.6 Work and coordinating size

Brick closers	These are cut bricks. They are visible on the face of the wall. An ordinary closer has dimensions of 46.25 × 102.5 × 65 mm and the queen closer is 46.25 × 65 × 215 mm.

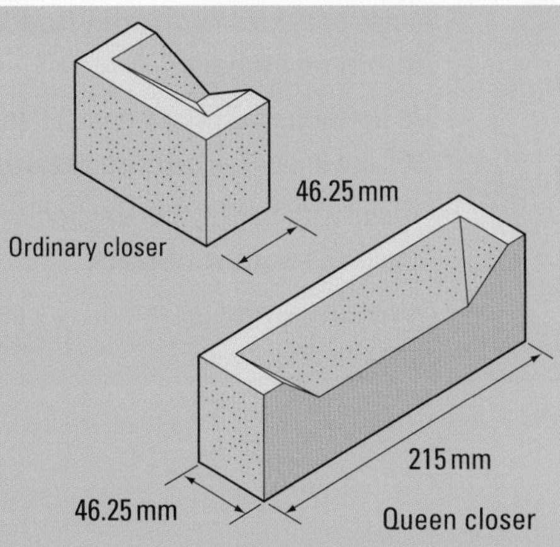

Figure 5.7 Ordinary closer and queen closer

Batts	Various terms are used to describe parts of the brick. A **half batt** is a brick that has been cut in half across the width, with dimensions of 102.5 × 102.5 × 65 mm. A **three-quarter batt** has dimensions of 158.75 × 102.5 × 65 mm.

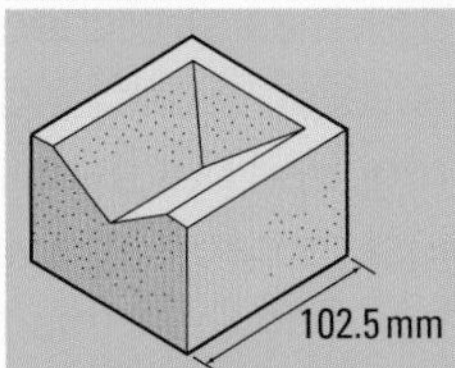

Figure 5.8 Half batt 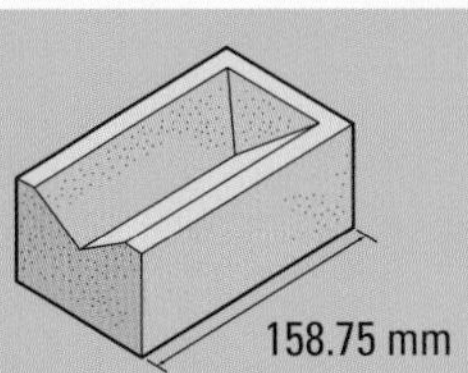   Figure 5.9 Three-quarter batt

Term	Explanation	
Bed joints and courses	Bed joints are the horizontal joints of mortar between each course of brick. A **course** is the term that is used to describe a row, usually horizontal, of bricks.	
Cross, transverse and wall joints	A **cross joint** is the vertical joint between two bricks. A cross joint is also known as a **perpend** or perp. The **transverse joint** is also known as a **sectional joint**. These are at right angles to the face of the wall. **Wall joints** are those that run parallel to the face of the wall.	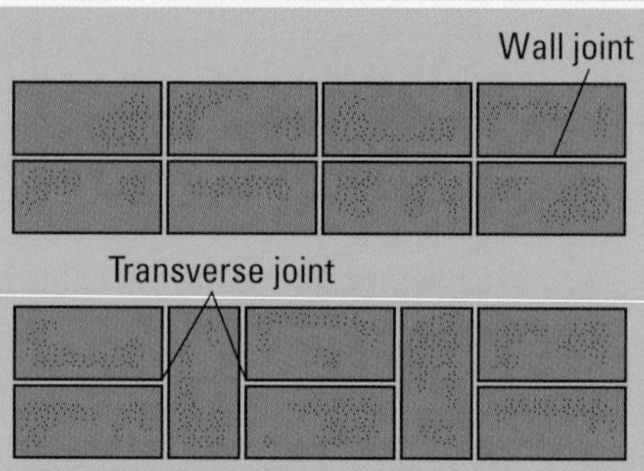 Figure 5.10 Wall and transverse joints
T-junctions and cross junctions	A **T-junction** is where two walls meet, with a T-shaped formation. A **cross junction** is where two walls cross over one another. Where a wall finishes and has a wall face it is known as a **stopped end**.	 Figure 5.11 T-junctions  Figure 5.12 Cross junctions
Quoins and return angles	On the face side of a wall corner or angle, depending on which direction you are looking at the wall, the bricks that form the angle are known as **quoin headers** or **quoin stretchers**. At junctions the bricks forming angles of 90° are known as **square quoins**. Those with angles that are not 90° are known as **squint quoins**. A **return angle** (or return) is a junction between two walls.	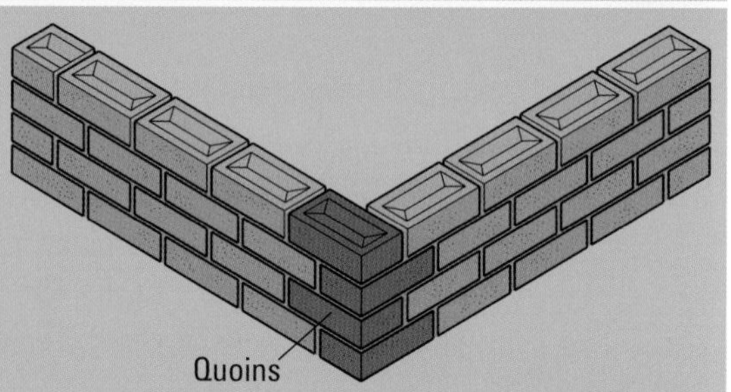 Figure 5.13 Return angle

Table 5.4 Brick terms

The simplest type of bonded wall is one where the bricks are all laid down as stretchers and they overlap one another by half a length (this is called a half brick wall). This would be fine if only a half brick wall was required but usually a one brick wall is needed. It would not be suitable just to cement two half brick walls together. They require better bonding than this to make them stable.

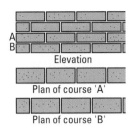

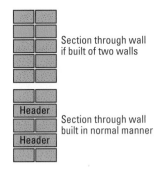

Section through wall
if built of two walls

Section through wall
built in normal manner

Figure 5.14 Half brick wall in plan and elevation

Figure 5.15 Two half brick walls side by side and a bonded wall

There has to be some kind of bonding across the two walls. This ensures that the load is distributed along the thickness and the length of the wall. If the length of the course of brickwork is not precisely the multiple of several brick lengths then batts would be inserted. This is known as a broken bond.

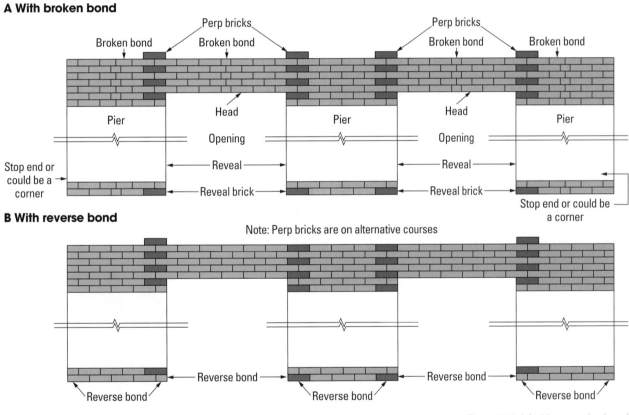

Figure 5.16 Marking out the bond

Dealing with broken bonds requires some planning. The position of the batt is not really important below ground level, but once ground level has been reached it is important to:

* accurately mark the position of any opening

* ensure that the broken bond is placed at the centre of any pier or opening (this ensures that all the vertical joints will be plumb and will also give a better appearance to the finished wall).

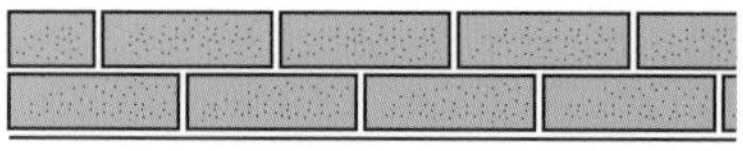

Reverse bonds

It is sometimes the practice to finish off a course of bricks with alternating end bricks, as can be seen in Fig 5.17.

Figure 5.17 Reverse bond

Whatever technique is used, the bonding must be systematic – it should follow a pattern. Bonding arrangements should be constant across the wall. Figure 5.18 shows corner bricks that are balanced. Figure 5.19 shows another length of wall, but this time the direction of the corner bricks has been reversed.

Stretcher   Stretcher

Figure 5.18 A length of wall with balanced corner bricks for each course

Stretcher   Header

Figure 5.19 Reverse bond

Transferring wall positions onto foundation concrete

Once the foundation concrete has set, ranging lines should be fixed at the wall lines on the profiles. A spirit level needs to be plumbed from the ranging line and then its position marked into mortar screed that has been spread onto the concrete. A mark needs to be made at the end of each wall to mark each corner.

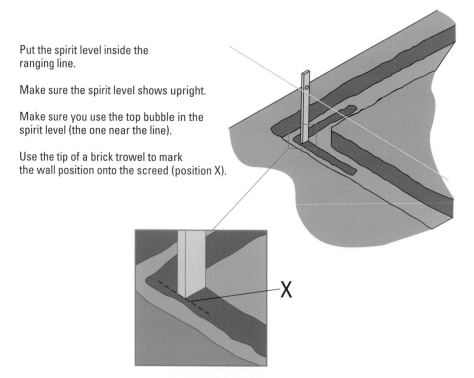

Put the spirit level inside the ranging line.

Make sure the spirit level shows upright.

Make sure you use the top bubble in the spirit level (the one near the line).

Use the tip of a brick trowel to mark the wall position onto the screed (position X).

Figure 5.20 Marking out screed from ranging lines

CASE STUDY

You're learning all the time

Gary Kirsop, Head of Property Services at South Tyneside Homes, says:

'Building solid walling applies especially to large retaining walls in gardens, for instance. The main thing is making sure that your solid walls are right, depending on what bond you're building them in, and what structure you're building them in. Solid walls are sometimes needed for structural purposes; some could be for retaining walls, retaining structures back or for support.

Keep in mind that there are different elements and designs, and you will need awareness of the range of all of them. The best way to get this awareness is from listening to your lecturers in college. You don't want to turn up to the site and be faced with something you've never experienced before. But, even if you do find yourself working on an unfamiliar structure, then don't be frightened to ask. If you're a young person who's learning and training and you don't know something yet, that's OK, but don't just fumble your way through it – ask. If you don't ask and go ahead to try to have a go yourself, then you're potentially costing money, straight away.

Learning skills are something you use throughout your apprenticeship, but also throughout your life. To me, it's been a long learning curve, and I'm still learning – we're all still learning.'

BUILDING SOLID WALLING TO THE GIVEN SPECIFICATION

This part of the chapter is best tackled through practicals, which give you the opportunity to learn all of the steps necessary to construct buildings successfully using solid walling. See also pages 233–4 for information about jointing and pointing.

Constructing walling to given datum heights

See Chapter 4 for information about setting out and datum points.

English bond is the strongest bond used in bricklaying as there are no straight joints in the wall. It is used in walls that are at least one brick thick.

English bond consists of alternate courses of headers and stretchers, with a queen closer next to the corner, producing a quarter bond. Some people think that English bond looks a bit boring and monotonous but it can be enhanced by using contrasting header.

KEY TERMS

English bond

– a one brick thick wall, laid to quarter lap, where the courses are alternate stretchers and headers.

Dry bonding is the term used for laying each brick in the first course without using any mortar joints either for the bed joint or the cross joint. Bricklayers need to do this in order to check that each course will work with exact brick sizes with no opening up or tightening up the cross joints needed or to check whether cut bricks will be needed to complete each course. It is also good practice to do this if you have any window or door openings to consider higher up. This enables you to set out the bonding with these in mind.

PRACTICAL TASK

1. BUILD AN ENGLISH BOND RETURN CORNER

OBJECTIVE

To complete an English bond return corner to the required standards.

Ensure you select PPE appropriate to the job and site where you are working. Refer to the PPE section in Chapter 1.

The model is to be built in local facing bricks.

The face side (stretchers) = 890 mm long

The return side (headers) = 740 mm

Joint finish

Front = weather struck

Rear = flush from the trowel

TOOLS AND EQUIPMENT

Walling trowel   Jointing iron

Pointing trowel   Builder's square

Spirit level   Bat gauge

Lump hammer and bolster chisel

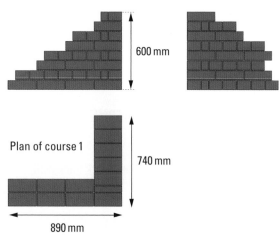

Plan of course 1

600 mm

740 mm

890 mm

Figure 5.21 English bond return corner

STEP 1 Work out the amount of materials that you will need.

Example

There are 120 bricks per square metre in a one brick wall. To find out how many bricks you will need, you have to find out how many square metres in the wall:

Length of wall (A) × Height of wall (B)

A × B = square metre (C)

You now have to multiply the square metres by 120

C × 120 = N (number of bricks)

Attention to the following points helps to maintain a good standard of work throughout all these practical exercises:

• Cut queen closers neatly and keep them regular in size.

• Remove mortar from the back of the bricks against the collar joint as it can prevent the backing up from being laid level.

• Keep perpends uniform and plumb because if the cross joints become too big then you will soon lose the bond, especially on the header course, and it will cause straight joints on the face of the wall.

• When laying the back of the wall, avoid using too much mortar near the collar joint as it could prevent the bricks from being laid in line. When you tap it back in place it will cause the face bricks to move out in front of the line.

Don't forget that you might break or damage some of the bricks so you need to add some extra for wastage – normally 5%

W (wastage) = N + 5% = Total number of bricks required

STEP 2 Cut seven queen closers to the correct size using the appropriate tools.

PRACTICAL TIP

Queen closers are cut down the length of a brick. It is important to cut the closers to the correct size. If they are too big then it will throw out the bond on the header course and you will then lose the bond and end up with straight joints. Do not just cut the brick down the middle as this will make the closer too wide and will make it difficult to keep the bond correct.

STEP 3 Set out a return corner (90° angle) from the face line of the first wall set out using a builder's square and check it using the 3:4:5 method for accuracy (see pages 110–11).

STEP 4 Lay the corner brick on the stretcher face to gauge and level both ways. Lay the remaining three bricks along the stretcher face, levelling them with the corner brick.

PRACTICAL TIP

When building return corners, don't forget that when you change direction you change bond so that on one face of the wall you will see stretchers and on the other face you will see headers.

STEP 5 Lay the header course, starting with a queen closer next to the corner brick. Check the corner with the builder's square for accuracy.

Make sure that the queen closer is not tilting (that it is level in its length and width).

At this point it is best to check that you haven't made the joints between the headers too big. Dry bond the stretchers along the top of the header course and make sure that the joint is in the centre of the header below. Make any adjustments you need to.

STEP 6 Complete the rest of the corner. Joint the wall as the work proceeds.

Flemish bond is another type of bond used for one brick walls or above. It consists of alternate stretchers and headers in the same course. The header should be placed in the centre of the stretcher on the course below and above it. This creates a more attractive bond than English bond.

The bond is achieved by introducing a queen closer next to the corner header, giving you a quarter bond.

PRACTICAL TASK

2. BUILD A FLEMISH BOND RETURN CORNER

OBJECTIVE

To complete a Flemish bond return corner to the required standards.

The model is to be built in local facing bricks.

Face = weather struck

Rear = flush finish as work proceeds

Ensure you select PPE appropriate to the job and site where you are working. Refer to the PPE section in Chapter 1.

TOOLS AND EQUIPMENT

Walling trowel   Builder's square

Pointing trowel   Jointing iron

Spirit level   Bat gauge

Lump hammer and bolster chisel

STEP 1 Work out the amount of materials that you would need to complete the model. See **Step 1** on page 145 (English bond return corner) for guidance on how to do this.

STEP 2 Cut the queen closers to the correct size, using the appropriate tools and wearing the right PPE. See the practical tip in **Step 2** on page 145 (English bond return corner) for guidance on how to do this.

STEP 3 Using a builder's square, set out a 90° angle and check it using the 3:4:5 method for accuracy. See pages 110–11 to remind yourself how to do this.

STEP 4 Lay the corner stretcher brick to gauge and level both ways. Lay the remaining bricks along the face, levelling them with the corner brick.

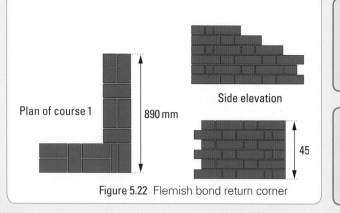

Plan of course 1   890 mm

Side elevation

45

Figure 5.22 Flemish bond return corner

STEP 5 Lay the return course, starting with a queen closer next to the corner brick. Check the corner with the builder's square for accuracy.

PRACTICAL TIP

At this point it is best to check that you have not made the joints between the bricks too big. Dry bond the second course along the top of the first course and make sure that the header is in the centre of the stretcher below. Make any adjustments at this point.

STEP 6 Complete the rest of the corner. Joint the wall as the work proceeds.

Figure 5.23 The completed corner

Sometimes you need to build an intersecting wall from the main wall. This is called a T-junction. Depending on the length of the T-junction, you can either build it in as you go or leave toothings to add the T-junction in at a later date. In either case, care must be taken to use the correct bonding arrangement at the T-junction.

PRACTICAL TASK

3. BUILD AN ENGLISH BOND WALL WITH A T-JUNCTION

OBJECTIVE

To complete an English bond wall with a T-junction to the required standards. The wall will include a raking cut and brick on edge.

TOOLS AND EQUIPMENT

Walling trowel	Jointing iron
Pointing trowel	Adjustable bevel
Spirit level	Corner blocks and line
Lump hammer and bolster chisel	Bat gauge
	10 mm timber packing × 2
Builder's square	

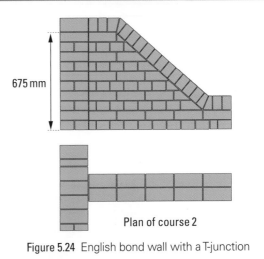

675 mm

Plan of course 2

Figure 5.24 English bond wall with a T-junction

Ensure you select PPE appropriate to the job and site where you are working. Refer to the PPE section in Chapter 1.

STEP 1 Work out the amount of materials that you would need to complete the model. See **Step 1** on page 145 (English bond return corner) for guidance on how to do this.

STEP 2 Cut the correct amount of queen closers to the correct size using the appropriate tools. See the practical tip in **Step 2** on page 145 (Build an English bond return corner) for guidance on how to cut queen closers

STEP 3 Dry bond the first course to the length shown on the drawing, ensuring that the T-junction is at 90°. See page 144 for guidance on dry bonding. Check the length of the wall.

PRACTICAL TIP

It is important that the junction wall is the correct size, otherwise the stretcher course will not fit.

STEP 4 Lay the first brick at one end to gauge and level it. Now lay the second end brick.

Using a straight edge and level, level the second end brick with the first.

STEP 5 Build small corners at each end of the wall.

STEP 6 Run in the bricks between the corners and complete the remaining wall, using toothing (see page 146) at the end where necessary. Joint the wall as the work proceeds.

Figure 5.25 Building up the wall with toothing

PRACTICAL TIP

Make sure that you keep the toothing clean and level or you will not fit the bricks into it.

STEP 7 Set up a line and complete the raking out.

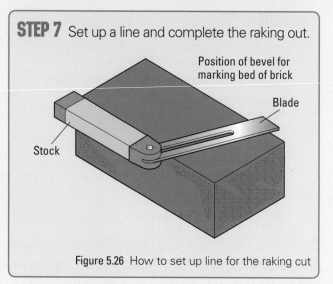

Position of bevel for marking bed of brick

Blade

Stock

Figure 5.26 How to set up line for the raking cut

PRACTICAL TIP

Raking cuts mean cutting the bricks at an angle, as seen on the gable walls of houses or on boundary walls where there is a difference in levels on the wall.

Make sure that you cut the bricks accurately and that they are level across the wall with no lumps. Otherwise when you are laying the brick on the edge you could end up with more than a 10 mm joint showing at the edges.

STEP 8 There are two methods of marking out the angle of rake on the brick.

The first method is using a 'chalk line': using a piece of chalk or chalk dust (cement can be used instead but take care with contact with skin). Deposit the chalk onto the line. The line needs to be fixed at two predetermined points on the wall: the starting point and finish point of the raking cut. As each course is laid, the line can be transferred on the face of the brick.

Using two 10 mm timber packing (one for the bed joint and one for the cross joint), place the brick.

Pull the line taut and let go. This will leave chalk onto the face of the brick. It can now be cut to the correct angle.

Figure 5.27 Marking out bricks for a raking cut, using a chalk line

Figure 5.28 Brick marked with chalk

The second method is to use an adjustable bevel. After setting up the line for the raking cut, set the adjustable bevel to the angle of the line. You can now mark the bricks using the adjustable bevel and cut them accurately.

Figure 5.29 Marking out bricks for a raking cut, using an adjustable bevel (1)

Figure 5.30 Marking out bricks for a raking cut, using an adjustable bevel (2)

STEP 9 Cut all the bricks to complete the rake, then lay the brick up the rake using the line as a guide.

PRACTICAL TIP

Raking cuts are best done by cutting off the surplus part of the brick then cutting the angle. If the cut angle is less than half a brick, it is best to cut the angle of a full brick then cut the brick to the correct length. Do not try to cut the sharp angle out of the frog side as this will normally crack and break off.

Do not lay any bricks above the line as you will not be able to lay the brick on edge.

Figure 5.31 Wrong method of cutting brick at an angle

STEP 10 To cut the angle bricks for the brick-on-edge, mark out the angle of the raking cut on a sheet of paper or a clean area on the floor.

Figure 5.32 Marking out the angle for the cuts

Using the drawing, mark out and cut two bricks for the bottom for the angle on the raking cut.

PRACTICAL TIP

The angle at the top of the raking cut should be the same as at the bottom.

STEP 11 Starting at the bottom lay the brick on edge, then joint the brick-on-edge on completion.

PRACTICAL TIP

Keep an eye on the gauge as you lay the brick on edge up the rake.

PRACTICAL TASK

4. BUILD A FLEMISH BOND T-JUNCTION

OBJECTIVE

To complete a section of a Flemish bond T-junction wall to the required standards.

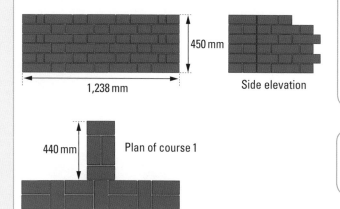

450 mm

1,238 mm

Side elevation

440 mm   Plan of course 1

Figure 5.33 Flemish bond T-junction

TOOLS AND EQUIPMENT

Walling trowel

Pointing trowel

Spirit level

Lump hammer and bolster chisel

Jointing iron

Builder's square

Bat gauge

Ensure you select PPE appropriate to the job and site where you are working. Refer to the PPE section in Chapter 1.

STEP 1 Work out the amount of materials that you would need to complete the model.

STEP 2 Cut the queen closers to the correct size, using the appropriate tools and wearing the right PPE.

STEP 3 Set out and dry bond the first course, ensuring that the T-junction is at 90°. Re-check the measurements.

STEP 4 Lay the first course and check it for accuracy.

STEP 5 Complete the remaining wall, using toothing on the 'T' junction.

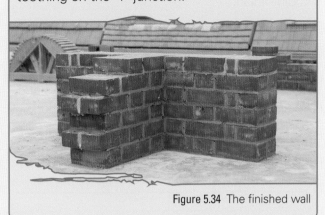

Figure 5.34 The finished wall

English garden wall bond

If you build a garden wall that can be seen on both sides it is not a good idea to use English bond. You cannot get a good finish on both sides of the wall due to the different sizes of bricks used in the header course. It is therefore best to use English garden wall bond.

English garden wall bond can either have three, five or even seven courses of stretcher courses to one course of headers. It is not as strong as English bond. The bond is maintained by introducing queen closers or a three-quarter batt at the quoin.

PRACTICAL TASK

5. BUILD A ONE-BRICK WALL IN ENGLISH GARDEN WALL BOND

OBJECTIVE

To complete an English garden wall to the required standards. The wall will include a decorative feature in contrasting bricks, oversail and brick-on-edge capping.

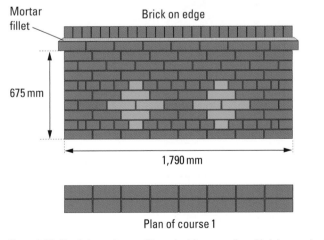

Mortar fillet

Brick on edge

675 mm

1,790 mm

Plan of course 1

Figure 5.35 English garden wall bond with oversail and brick-on-edge

The model is to be built in local facing bricks.

Joint finish:

Front = half round

Rear = flush from the trowel

Shades are to be built in contrasting bricks and pointed with coloured mortar on completion.

Ensure you select PPE appropriate to the job and site where you are working. Refer to the PPE section in Chapter 1.

TOOLS AND EQUIPMENT

Walling trowel	Jointing iron
Pointing trowel	Corner blocks and line
Spirit level	Straight edge
Lump hammer and bolster chisel	Bat gauge

STEP 1 Work out the amount of materials that you would need to complete the model.

STEP 2 Cut the queen closers to the correct size, using the appropriate tools and wearing the right PPE.

STEP 3 Set out and dry bond the first course.

Lay the first brick at one end to gauge and level. Now lay the second end brick at the correct distance from the first one.

Using a straight edge and level, level the second end brick with the first.

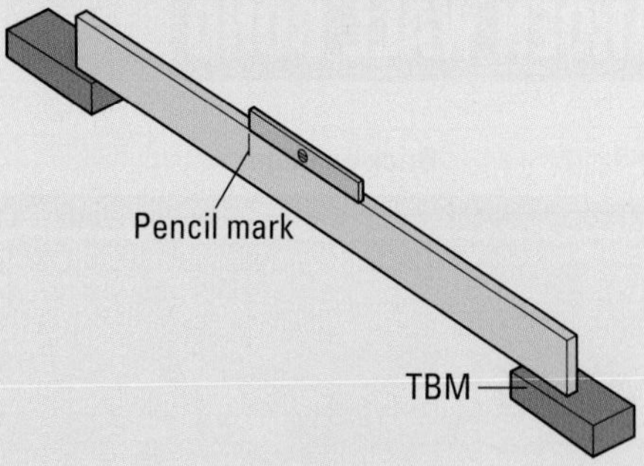

Pencil mark

TBM

Figure 5.36 Transferring levels with straight edge and level

STEP 4 Starting with a header course, build a small corner at each end.

PRACTICAL TIP

Remember you are building three courses of stretchers to one course of headers.

STEP 5 Wall the bricks in between the corners and don't forget to use the contrasting bricks for the pattern.

STEP 6 Build two more small corners and wall in between.

PRACTICAL TIP

Don't forget to use the contrasting bricks.

STEP 7 Re-point the contrasting bricks with a coloured mortar.

STEP 8 Lay the oversail course.

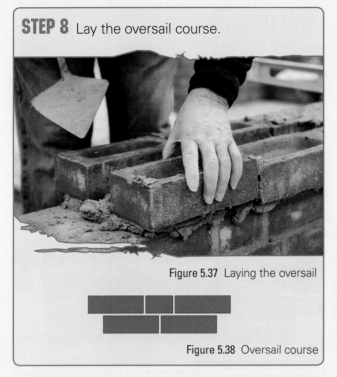

Figure 5.37 Laying the oversail

Figure 5.38 Oversail course

PRACTICAL TIP

Oversail courses are used to protect the top of the wall from the weather. Usually you would use a water-resistant brick for the brick-on-edge as you don't want any water to penetrate the top of the wall. Along with a mortar fillet it allows rainwater to run off the top of the wall and away from the face of the main wall.

PRACTICAL TIP

Make sure the projection is the correct size, and that none of the bricks are tipping.

STEP 9 Lay half a block at each end of the wall. If you start laying the brick-on-edge at the end of the wall, the end brick will always move when you are laying the next brick. By laying half a block on each end you have something solid to push the brick against.

PRACTICAL TIP

Although the bricklayer in the photos is not wearing gloves for clarity, you should always wear appropriate PPE, including hi vis.

Figure 5.39 Block on end of wall

Figure 5.40 Placing the first brick against the block

Figure 5.41 Filling joints

Figure 5.42 Placing second brick

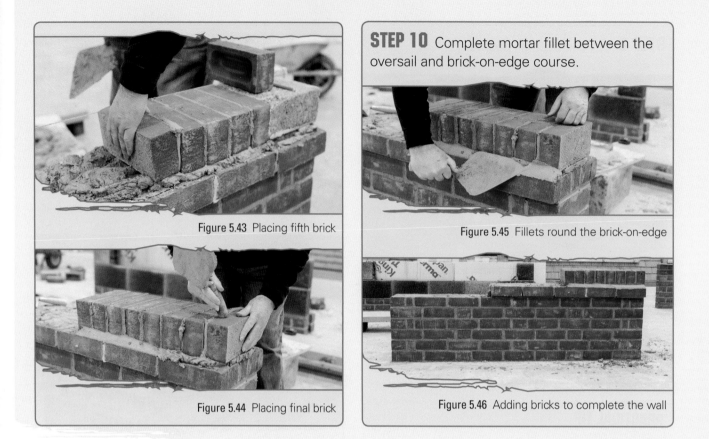

Figure 5.43 Placing fifth brick

STEP 10 Complete mortar fillet between the oversail and brick-on-edge course.

Figure 5.45 Fillets round the brick-on-edge

Figure 5.44 Placing final brick

Figure 5.46 Adding bricks to complete the wall

Flemish garden wall bond

Flemish garden wall bond differs from English garden wall bond by having three stretchers to one header in the same course. The header should be placed in the centre of the middle stretcher of the course below.

As in all single brick walls, the bond is maintained by using a queen closer next to the corner header.

PRACTICAL TASK

6. BUILD A SINGLE BRICK WALL IN FLEMISH GARDEN WALL BOND

OBJECTIVE

To complete a Flemish garden wall bond to the required standards.

The model is to be built in local facing bricks. Rear of wall to be flush finish as work proceeds.

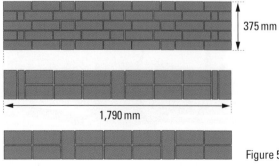

375 mm

1,790 mm

Figure 5.47 Plan of Flemish garden wall bond wall

TOOLS AND EQUIPMENT

Walling trowel	Builder's square
Pointing trowel	Jointing iron
Spirit level	Straight edge
Lump hammer and bolster chisel	Bat gauge

Ensure you select PPE appropriate to the job and site where you are working. Refer to the PPE section in Chapter 1.

STEP 1 Work out the amount of materials that you would need to complete the model.

STEP 2 Cut the queen closers to the correct size, using the appropriate tools and wearing the right PPE.

STEP 3 Dry bond the first course. Re-check the dimensions.

STEP 4 Lay the first brick at one end to gauge and level. Now lay the second end brick.

Using a straight edge and level, level the second end brick with the first.

STEP 5 Build two small corners at each end. Make sure that you keep the garden wall bond. If you do not keep the bond then you will not be able to fit all the bricks into the wall.

STEP 6 Using corner blocks and line run in the courses between the corners. Joint the wall as the work proceeds. Build the wall to the required dimensions.

Figure 5.48 This photo shows a combination of Flemish garden wall bond (top band of bricks above coloured course) and English garden wall bond (below coloured course)

REED TIP

Employers will want to know that you understand the importance of health and safety. Make sure you know the reasons for each safe working practice.

Damp-proof barriers

See also the cavity wall chapter.

A horizontal damp-proof course must be inserted 150 mm above ground level. Other considerations are as follows:

* Wherever there is a join, such as at a junction of a wall, there should be a minimum overlap of the damp-proof barrier of 100 mm.

* For two courses above ground level and one below frost resistant bricks should be used.

* The face of brickwork on sloping sites should be waterproofed.

* Although most bricks can be used below the damp-proof course, mortar needs to be sulphate resisting.

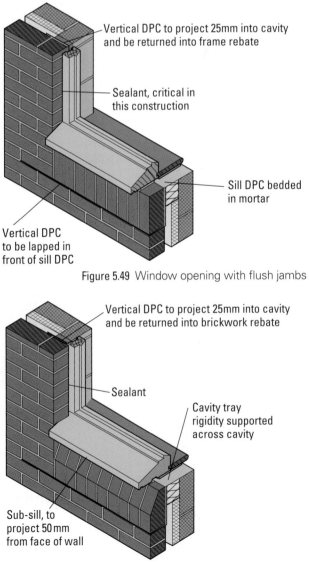

Vertical DPC to project 25mm into cavity and be returned into frame rebate

Sealant, critical in this construction

Sill DPC bedded in mortar

Vertical DPC to be lapped in front of sill DPC

Figure 5.49 Window opening with flush jambs

Vertical DPC to project 25mm into cavity and be returned into brickwork rebate

Sealant

Cavity tray rigidity supported across cavity

Sub-sill, to project 50 mm from face of wall

Figure 5.50 Window with rebated jambs

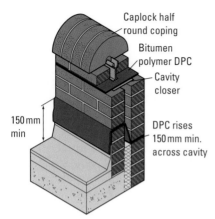

Figure 5.51 Parapet wall

Caplock half round coping
Bitumen polymer DPC
Cavity closer
DPC rises 150mm min. across cavity
150mm min

As the name suggests, a damp-proof barrier aims to protect the building against dampness. The following safeguards must always be met:

* Damp-proof barriers should be flexible and only bedded into smooth mortar.

* Any exposed edges should not have mortar or render left on them.

* The damp-proof barrier should not extend into a cavity.

* Specific jobs require specific damp-proof barriers so the correct one always needs to be used.

* No stones or any other debris should be allowed to get into the mortar, as this may puncture the membrane.

There are several different types of damp-proof barrier, which are outlined in Table 5.5.

Type of barrier	Description and properties
Lead core reinforced	Consist of a core of lead, which is reinforced with hessian and then coated with a polymer. The material has a finish of silica sand. It can be used for all different types of masonry wall.
Pitch polymer	A high performance material that can be used in all different types of walling. It is sometimes used alongside self-adhesive waterproof membranes.
Self-adhesive	Generally used for tanking, mainly below ground level or for solid concrete floors.
Engineering bricks	Bedded into cement mortar and well bonded to produce a waterproof surface.

Table 5.5 Different types of damp-proof barrier

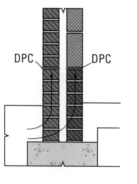

DPC   DPC

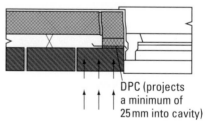

DPC (projects a minimum of 25mm into cavity)

Figure 5.52 Rising and horizontal damp protection

KEY TERMS

Tanking

– this is the process that involves waterproofing horizontally and vertically. It is common to tank basements where waterproofing is essential to protect the dwelling.

Maintaining industrial standards

Industrial standards means following clearly accepted guidelines and quality of work. Organisations such as the Concrete Block Association offer sound advice regarding good practice.

Building squint corners

Not all corners are 90° – in some cases they have to be built at an angle, for example to follow the boundaries of the site or for bay windows. They can be either acute (less than 90°) or obtuse (more than 90°).

In most cases, any angled work uses specially shaped, purpose-made bricks. These are known as 'squints'.

Building squint corners requires more skill than building basic straight walls. It is important to keep the corner plumb and to gauge it carefully. You can use a squint brick to obtain the correct bond, or you can use a specially shaped brick to overcome this and maintain the bond.

PRACTICAL TASK

7. BUILD A SECTION OF ENGLISH BOND WITH AN OBTUSE (SQUINT) CORNER

OBJECTIVE

To build a squint corner in English bond to the required standards.

The model is to be built in local facing bricks.

The joint finish for the front is weather struck. The rear is flush from the trowel.

TOOLS AND EQUIPMENT

Walling trowel	Jointing iron
Pointing trowel	Straight edge
Spirit level	Corner blocks and line
Bat gauge	Template
Lump hammer and bolster chisel	

Ensure you select PPE appropriate to the job and site where you are working. Refer to the PPE section in Chapter 1.

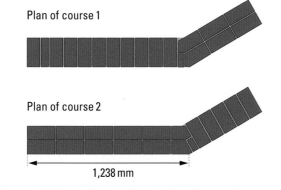

Plan of course 1

Plan of course 2

1,238 mm

Figure 5.53 Plan of English bond wall with a squint corner

STEP 1 Work out the amount of materials that you would need to complete the model.

STEP 2 Cut the queen closers to the correct size, using the appropriate tools and wearing the right PPE.

STEP 3 Set out a 45° angle and make a template to the correct angle.

STEP 4 Set out and dry bond the first course. Make sure that the header course is set out properly with the correct size joints – check it by laying stretchers along the top.

STEP 5 Lay the first course and use the template to check that the angle is correct. Ask your tutor to check it and to point out the plumbing points.

Figure 5.54 Checking the angle and plumbing points

STEP 6 Continue to build the wall until you have reached the correct height. Joint the wall as the work proceeds and note the position of the plumbing points.

Figure 5.55 The finished wall from one side…

Figure 5.56 …and the other

Figure 5.57 Mesh reinforcing the brickwork

Reinforcement

Movement can cause cracking but this can be offset by bed joint reinforcement.

A common method is to add in mesh. These are made from steel and are used to help prevent cracking. They are designed to fit in with the sizes of common bricks and blockwork. These mesh sections go in horizontally between courses.

Coping

Coping is a form of capping or covering that is placed on the top of a wall. The coping can either slope in a single direction or in two directions. It tends to be made of brick or concrete.

Vertical movement joints

Over the years it has been recognised that walls over 10 m long are prone to cracking due to movement. The causes of the movement are temperature changes and the natural processes of the wall becoming wet and then drying out.

In order to overcome this problem, long lengths of wall are effectively split up into shorter lengths. Between each length is a filler, which can be compressed and allows a certain amount of movement. The filler is covered in mastic in order to seal it. It can still move, but it is protected from water penetration.

The manufacturers of blocks suggest that the first movement joint should be within 3 m of a corner or a fixed end. They also suggest that walling lengths should be no longer than 6 m with vertical movement joints placed in between them.

The construction of the movement joints can be achieved by using ties, which can effectively connect the lengths of wall together but still allow movement. As an alternative, a 10 mm gap can be filled with fibre board.

Figure 5.58 Mixing mortar

Mixing jointing compounds

When mixing mortar, it is important to make sure that it is strong enough and is workable. Therefore there are no fixed proportions for mixing. Some sands need more cement or lime than others. For lime mortar, the general rule of thumb is one part lime to three parts sand by volume.

The usual way of mixing is to put the cement on one side and then spread it over the sand, mixing the two together with a shovel. Once the two have been thoroughly mixed, a ring is created and then water is added.

It is very difficult to ensure that the same mix of mortar is being produced each time. The shovel is not an accurate measuring tool. There are two ways of getting around this:

* For larger jobs a method known as weight batching can be used. As the name suggests, this means using mixers with automatic hoppers. They drop the correct amount of material into the machine.

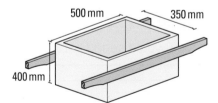

Figure 5.59 Gauge box

* For smaller jobs a bucket can be used. Some builders prefer to use gauge boxes. The dry materials are put inside the box, which does not have a bottom. You dump the materials into the box to get the proportions right (by eye and experience), and when you are happy with the proportions you simply lift the box out of the way and are left with the pile of materials to mix up. In this way, the right proportions of materials can be measured out.

Task 13 on page 169 shows you how to build a vertical movement joint.

8. BUILD A SECTION OF ENGLISH BOND WITH CONCRETE COPING

OBJECTIVE

To complete a section of an English bond wall to the required standards. The wall will include a concrete coping to the top of the wall.

The model is to be built in local facing bricks.

Joint finish

Front = weather struck

Rear = flush from the trowel

The joints on the concrete coping should be left flush.

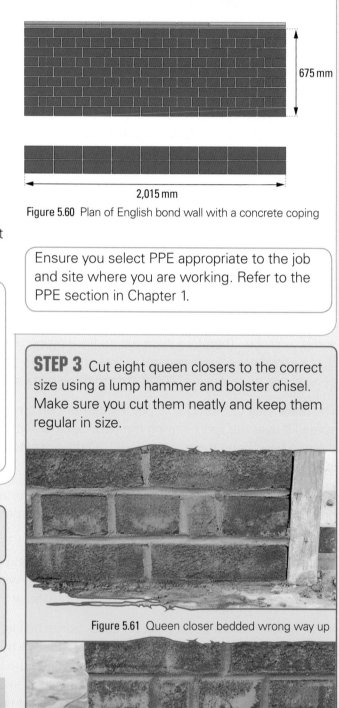

Figure 5.60 Plan of English bond wall with a concrete coping

Ensure you select PPE appropriate to the job and site where you are working. Refer to the PPE section in Chapter 1.

TOOLS AND EQUIPMENT

Walling trowel	Jointing iron
Pointing trowel	Straight edge
Spirit level	Bat gauge

Lump hammer and bolster chisel

Corner blocks and line

STEP 1 Work out the amount of materials that you would need to complete the model.

STEP 2 Set out and dry bond the first course to the correct length of the wall as shown on the drawing. Check the length of the wall.

STEP 3 Cut eight queen closers to the correct size using a lump hammer and bolster chisel. Make sure you cut them neatly and keep them regular in size.

Figure 5.61 Queen closer bedded wrong way up

PRACTICAL TIP

Always lay the queen closers with the full face of the bed of the brick down. If you lay the closers 'frog' down you only have three edges of the frog on the brick, leaving one edge with no support and the closer will tend to tilt, which will show up in the wall.

Figure 5.62 Queen closer bedded right way up

STEP 4 Lay the first end brick at one end to gauge and level. Now lay the second end brick at the other end, the correct distance from the first.

PRACTICAL TIP

Always gauge the brick first; if you level it and it is not to gauge then you will have to re-lay the brick, therefore wasting time and effort in levelling it first.

Using a straight edge and level, level the second end brick with the first.

PRACTICAL TIP

When working in foundations never assume that the foundation concrete is level. Always check it before you start laying bricks and always start at the highest point. It is easier to bed the bricks up with more mortar than cutting the bricks down lengthways.

STEP 5 Using corner blocks and a line lay the rest of the course on the face of the wall. When laying the bed joint for the back of the wall, always push any excess mortar away from the bricks already laid because at this point you need to check the width of the wall. You do this by laying a brick across the bricks that have been previously laid. If there is too much mortar between the two bricks (collar joint) then you will push out the face brick as you tap the rear brick to get the correct width.

PRACTICAL TIP

At this point the two corner bricks will still be 'wet' (not set) so you will have to put some bricks on top of the corner bricks so that they will not move.

Figure 5.63 Weighing down the corner bricks

Figure 5.64 Removing excess mortar from back of wall

Figure 5.65 Checking the width of the wall

STEP 6 Build two small corners (five courses) at each end. Joint the corners as the work proceeds. Make sure that you keep the bond, otherwise when you run the main wall (build the rest of the wall) you will not fit all the bricks in. Make sure that you place the header over the centre of the joint on the stretcher course below.

PRACTICAL TIP

It is a myth that building big corners is quicker. It is quicker to lay bricks to the line rather than having to keep picking up a spirit level to check the level of the bricks and to correct them.

STEP 7 Wall the bricks to a line between the corners. Joint the wall as the work proceeds.

STEP 8 Repeat **Steps 6** and **7** until the wall is the correct height.

STEP 9 Lay a concrete coping at each end of the wall. Make sure that the overhang is equal on both sides.

Figure 5.66 Laying coping at the end

PRACTICAL TIP

If there is a cut piece of coping, never put it at the end of the wall – always put it in the middle or next to any attached piers. A cut piece in a stretch of wall looks better in the middle rather than one end. It also has more resistance to movement or displacement due to its smaller size. Finally, it is easier for a small piece of coping at the end of a wall to get knocked and come loose.

STEP 10 Using corner blocks and lines, lay the rest of the copings along the wall. Check the finished course with the straight edge along the top to make sure that all the copings are level.

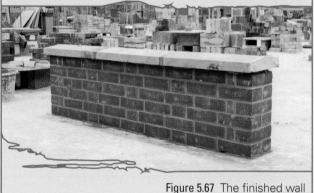

Figure 5.67 The finished wall

BUILDING ISOLATED AND ATTACHED PIERS TO THE GIVEN SPECIFICATION

A pier or pillar aims to give additional strength to a section of walling. It also makes the wall more stable. Piers are short lengths of walling bonded on every side. The smallest type of pier is usually 215 × 215 mm, which is two bricks laid side by side, as can be seen in Fig 7.69.

Piers that are larger than one brick can be bonded using either the English or Flemish bond, as can be seen in the subsequent diagrams.

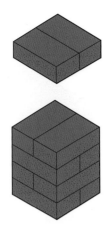
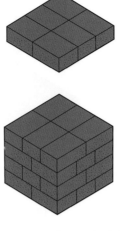
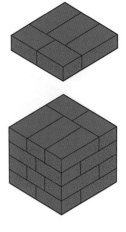

Figure 5.68 One brick pier

Figure 5.69 One and a half pier in English bond

Figure 5.70 Alternative arrangement

Figure 5.71 Two brick pier in Flemish bond

Figure 5.72 Two brick pier in English bond

Figure 5.73 Two and a half pier in Flemish bond

Figure 5.74 Two and a half pier in English bond

Bonds for isolated and attached piers

Attached piers are those that have a definite connection to a wall and create a T or a cross. These can be constructed using either English or Flemish bonds, as can be seen in the diagrams that follow.

Attached piers give lateral support to long free-standing walls. When building walls with attached piers, it is important to keep the same bond on the face of the pier as for the main wall.

Isolated (detached) piers can be either built as a decorative finish to a wall or to support a gate or iron railings in between the piers. Each side must be treated as the face of the wall. If they are to carry anything then they can be strengthened by filling the core with concrete.

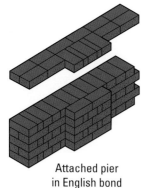
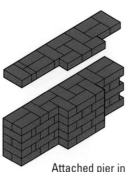

Attached pier
in English bond

Attached pier in
Flemish bond

Figure 5.75 Attached piers in English and Flemish bonds

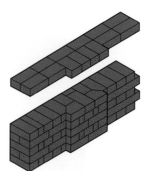
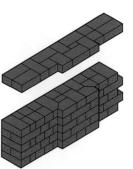

Attached pier
in English bond,
with smaller projection

Attached pier
in Flemish bond,
with smaller projection

Figure 5.76 Alternative attached piers in English and
Flemish bonds

PRACTICAL TASK

9. BUILD A SECTION OF ENGLISH BOND WALL WITH AN ATTACHED PIER

OBJECTIVE

To complete a section of one-brick wall with an attached pier to the required standards.

The model is to be built in local facing bricks.

Joint finish:

Front = weather struck

Rear = flush from the trowel

Front elevation

300 mm

2,250 mm

Plan of course 1

Figure 5.77 Plan of English bond wall with attached pier

TOOLS AND EQUIPMENT

Walling trowel

Builder's square

Pointing trowel

Jointing iron

Spirit level

Bat gauge

Lump hammer and bolster chisel

Corner blocks and line

Ensure you select PPE appropriate to the job and site where you are working. Refer to the PPE section in Chapter 1.

STEP 1 Work out the amount of materials that you would need to complete the model.

STEP 2 Cut the queen closers to the correct size, using the appropriate tools and PPE.

STEP 3 Set out and dry bond the first course, ensuring that the attached pier is at 90°.

STEP 4 Lay the first brick at one end to gauge and level. Now lay the second end brick.

Using a straight edge and level, level the second end brick with the first. Check measurement for length of wall.

STEP 5 Lay the first course.

STEP 6 Build a small corner at each end.

STEP 7 Complete the remaining wall. Joint the wall as the work proceeds.

PRACTICAL TASK

10. COMPLETE A SECTION OF FLEMISH BOND WALL WITH ATTACHED PIERS

OBJECTIVE

To complete a section of a Flemish bond wall to the required standards. The wall will include attached piers.

The model is to be built in local facing bricks.

Joint finish

Front = half round

Rear = flush from the trowel

TOOLS AND EQUIPMENT

Walling trowel	Builder's square
Pointing trowel	Jointing iron
Spirit level	Bat gauge
Lump hammer and bolster chisel	
Corner blocks and line	

Ensure you select PPE appropriate to the job and site where you are working. Refer to the PPE section in Chapter 1.

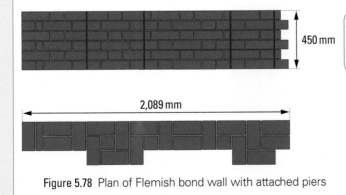

Figure 5.78 Plan of Flemish bond wall with attached piers

STEP 1 Work out the amount of materials that you would need to complete the model.

STEP 2 Cut the queen closers to the correct size, using the appropriate tools and wearing the right PPE.

STEP 3 Set out and dry bond the first course.

PRACTICAL TIP

There are two different types of attached piers to this wall. Take care in using the correct bonding arrangement.

STEP 4 Lay the first brick at one end to gauge and level. Now lay the second end brick.

Using a straight edge and level, level the second end brick with the first.

STEP 5 Complete the first course. Check the piers for square.

STEP 6 Build two corners at each end of the wall.

STEP 7 Complete the remaining wall. Joint the wall as the work proceeds.

PRACTICAL TIP

As you can only plumb one side of a small pier, it is better to select your bricks to ensure that they are all the same size. This makes the side that isn't plumbed look reasonably plumb.

Figure 5.79 The finished wall

Figure 5.80 Close up to show cut bricks

PRACTICAL TASK

11. BUILD A TWO-BRICK SOLID ISOLATED PIER IN ENGLISH BOND

OBJECTIVE

To build a two-brick solid isolated (detached) pier in English bond to the required standard.

The model is to be built in local facing bricks.

The wall should be finished with two sides weather struck joint as the work proceeds and two sides raked out and re-pointed half round on completion of the pier. The brick-on-edge capping should be finished with a flush joint.

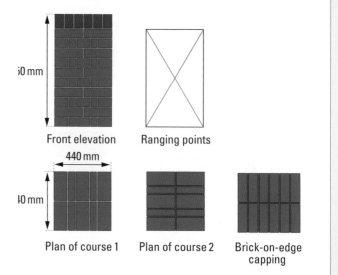

50 mm

Front elevation   Ranging points

440 mm

40 mm

Plan of course 1   Plan of course 2   Brick-on-edge capping

Figure 5.81 Plan of two-brick solid isolated pier in English bond

TOOLS AND EQUIPMENT

Walling trowel   Builder's square

Pointing trowel   Jointing iron

Spirit level   Bat gauge

Lump hammer and bolster chisel

Chariot (for raking out joints)

Ensure you select PPE appropriate to the job and site where you are working. Refer to the PPE section in Chapter 1.

STEP 1 Work out the amount of materials that you would need to complete the model.

STEP 2 Cut the correct amount of queen closers. It is important that you cut them to the correct size. If they are too big you will not be able to make the pier the right size.

STEP 3 Set out and dry bond the first course, making sure that it is a 440 mm square. Check the size and make sure it is square by checking the diagonals. Each diagonal should measure the same.

STEP 4 Lay the first course of bricks. Make sure that it is the correct size and that it is square.

PRACTICAL TIP

Before you carry on, try two stretchers across the header course and make sure that the joint in the middle is not more than 10 mm.

When you are laying each course, this is one of the few occasions when it is best not to use a full cross joint. If you put full cross joints on, tapping one side of the pier to get it plumb will move the other brick next to it, as there is no give between the two bricks. If the cross joint is not full then the mortar between the bricks will squeeze into the gap and the bricks will not move.

STEP 5 Continue to build the pier until it is the correct height. Every three courses, check the size and that it is square. Don't forget to joint the wall on two sides as you build the pier and rake out the other two sides.

STEP 6 Re-point the two raked out sides with a half-round joint finish.

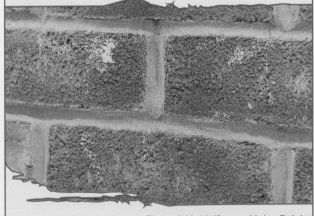

Figure 5.82 Half-round joint finish

STEP 7 Using an appropriate weather resisting brick, such as an engineering brick, build the brick-on-edge capping to the pier. Care must be taken when laying the end bricks as they can be easily disturbed and fall off.

PRACTICAL TIP

The top of the wall must be protected against the weather, otherwise water will get into the pier, corroding the bricks and making them brittle. In winter the water will freeze, making the bricks lift and come away.

PRACTICAL TASK

12. BUILD A TWO-BRICK HOLLOW ISOLATED PIER

OBJECTIVE

To build the hollow detached pier with oversail course shown below.

The core of the pier must be kept clear of any mortar.

The wall should be finished with two sides weather struck joint as the work proceeds and two sides raked out and repointed half round, on completion of the pier.

Ensure you select PPE appropriate to the job and site where you are working. Refer to the PPE section in Chapter 1.

TOOLS AND EQUIPMENT

Walling trowel   Builder's square

Pointing trowel   Jointing iron

Spirit level   Lump hammer and bolster chisel

Chariot (for raking out joints)

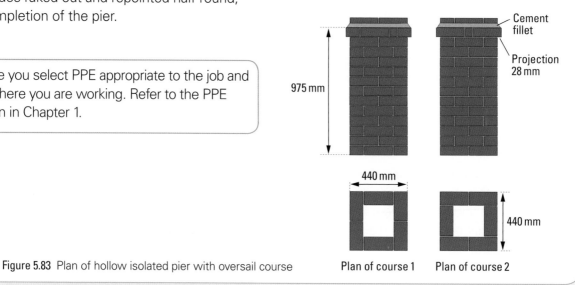

Figure 5.83 Plan of hollow isolated pier with oversail course

Plan of course 1   Plan of course 2

STEP 1 Work out the amount of materials that you would need to complete the model.

STEP 2 Set out a square measuring 440 mm on each side. Check the size and make sure it is square by checking the diagonals. Each diagonal should measure the same (616 mm).

Figure 5.84 Starting the pier

PRACTICAL TIP

If the square is more than 440 mm, it will be very obvious because the cross joint will be more than 10 mm.

STEP 3 If you are building several piers in a line, set out the two end piers and use corner blocks and line between the two piers.

You can now lay the other piers to the line. This makes sure that piers aren't sticking out too far, and that they're not twisted and out of line.

STEP 4 Lay the first course of bricks in stretcher bond. Make sure that it is the correct size and that it is square. Re-check the diagonals.

PRACTICAL TIP

When you are laying each course, this is one of the few occasions when it is best not to use a full cross joint.

If you put full cross joints on, tapping one side of the pier to get it plumb will move the other brick next to it, as there is no give between the two bricks. If the cross joint is not full then the mortar between the bricks will squeeze into the gap and the bricks will not move. Once the course is correct you can carefully fill in the cross joints.

STEP 6 Lay the oversail course (see Step 8 One-brick wall in English garden wall bond on page 153).

PRACTICAL TIP

Make sure that you get the projection the correct size, and that none of the bricks are tipping.

STEP 7 Lay the top course of bricks, set back so that they are the same size as the original pier.

STEP 5 Continue to build the pier to the correct height. Keep checking the size and that it is square every three courses. It is very easy to start getting out of square and building a twist in the pier. Remember to joint two sides with a weather struck joint as you are building the pier and rake out the other two sides to be re-pointed on completion.

STEP 8 Complete the fillet joint around the top of the oversail.

The fillet allows any rainwater to run off the top of the pier and prevent it from getting into the joint. If water was allowed to seep into the joint, in winter it would freeze and the brick would become loose.

Figure 5.85 Building up the pier

Figure 5.88 Filleting around oversail course

Figure 5.86 Setting out the oversail

STEP 9 Re-point the two raked out sides with a half round joint finish.

Figure 5.87 Levelling the oversail

Figure 5.89 The finished wall

13. BUILD A STRAIGHT FLEMISH BOND WALL WITH ATTACHED PIERS AND MOVEMENT JOINT

OBJECTIVE

To complete a section of a Flemish bond wall with attached piers, including a movement joint to the required standards.

The wall will include a concrete coping to the top of the wall.

The model is to be built in local facing bricks.

The expansion joint is to be built in as the work proceeds.

Joint finish

Front = half round

Rear = flush from the trowel

Ensure you select PPE appropriate to the job and site where you are working. Refer to the PPE section in Chapter 1.

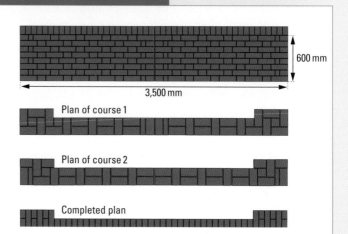

Figure 5.90 Plan of straight Flemish bond wall with attached piers and movement joint

TOOLS AND EQUIPMENT

Walling trowel

Builder's square

Pointing trowel

Jointing iron

Spirit level

Lump hammer and bolster chisel

Corner blocks and line

PRACTICAL TIP

This task has been designed as a two-person job. The wall can be reduced in size if one person does it.

PRACTICAL TIP

A movement joint is a continuous joint in brickwork, filled with a compressible material to allow for any movement in long continuous walls due to thermal or structural effects. Normally movement joints are placed every 10–12 m.

STEP 1 Work out the amount of materials that you would need to complete the model.

STEP 2 Cut the queen closers to the correct size, using the appropriate tools and wearing the right PPE.

STEP 3 Set out and dry bond the first course. Don't forget to leave a big enough gap for the movement joint (usually between 10 mm and 20 mm wide).

STEP 4 Lay the first brick at one end to gauge and level. Now lay the second end brick.

Using a straight edge and level, level the second end brick with the first.

STEP 5 Build two small corners (five courses) at each end. Make sure that you keep the bond. If you do not keep the bond then you will not be able to build the expansion joint in properly.

STEP 6 Using corner blocks and line, run in the courses between the corners.

STEP 7 Repeat **Steps 4** and **5** until you have reached the correct height.

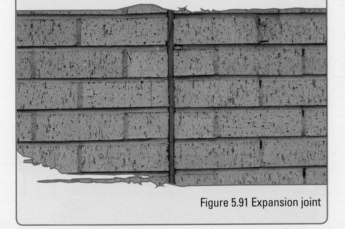

Figure 5.91 Expansion joint

STEP 8 Lay a block at each end of the wall. If you start laying the brick-on-edge at the end of the wall, the end brick will always move when you lay the next brick. By laying half a block on each end there is something solid to push the brick against.

STEP 9 Lay the brick-on-edge course between the blocks. The brick-on-edge should have a 3 mm fall across the length of the brick to allow any water to run off it. Some bricklayers use a line at each side of the brick-on-edge as a guide line, while others just use one line.

STEP 10 Remove the half blocks at each end and lay the remaining brick-on-edge.

PRACTICAL TASK

14. BUILD A HALF-BRICK WALL WITH ATTACHED PIER AND RAKING CUT

OBJECTIVE

To build a half-brick wall with an attached pier and raking cut to match the plan.

The model is to be built in local facing bricks. The raking cut should be built to industrial standards. The joint finish is face and pier half round and rear flush from the trowel. The core of the pier must be kept clear of any mortar.

There is another way of strengthening a half-brick wall by attaching a pier at the end of the wall where the gate will be. The raking cut finishes off a wall that has been extended to strengthen the pier.

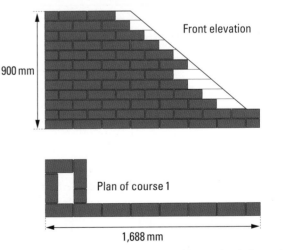

Figure 5.92 Plan of half-brick wall with attached pier and raking cut

Ensure you select PPE appropriate to the job and site where you are working. Refer to the PPE section in Chapter 1.

TOOLS AND EQUIPMENT

Walling trowel	Spirit level	Corner blocks and line
Builder's square	Adjustable bevel	10 mm timber packing × 2
Pointing trowel	Lump hammer and bolster chisel	Bat gauge
Jointing iron		

STEP 1 Work out the amount of materials that you would need to complete the model.

STEP 2 Set out and dry bond the first course as shown on the drawing. Re-check the length of the wall.

STEP 3 Lay the first brick at one end to gauge and level. Now lay the second end brick.

Using a straight edge and level, level the second end brick with the first.

Re-check the length of the wall and make sure the pier is square.

STEP 4 Complete the first course.

STEP 5 Build the shaded area of the model as shown on the plan.

PRACTICAL TIP

As the pier is not tied into the main wall you need to either use a wall tie or some reinforcing wire across the vertical joint in the corner to stop the pier coming away from the main wall.

When you do the toothing make sure it is level and clean on the underside because if you are using cement mortar it would set, meaning a brick could not be built in properly.

STEP 6 Set up a line for the raking cut. The line extends half a brick at the top and is set back half a brick at the bottom.

Figure 5.93 Before attaching a line

Figure 5.94 Measuring to attach a line for raking cut

STEP 7 Using either an adjustable bevel or a chalk line mark out the angle on the brick that need to be cut as shown. See **Step 7** on page 148 (Build an English bond wall with a T-junction) for guidance on how to do this.

Set the brick in position on 10 mm spacers, one for bed joint and one for cross joint.

Transfer the angle of the line onto the brick using a chalk line or adjustable bevel. If the bevel is used to mark the line of the raking cut, it can be used to extend pencil lines across the whole face of the brick.

Cut the brick to angle. Trim cut the surface of the brick with a comb hammer to ensure the core of the brick is not proud of the line. Bed the brick in position.

STEP 8 Cut the bricks to the correct angle.

PRACTICAL TIP

This is best done by cutting off the surplus part of the brick then cutting the angle. If the cut angle is less than half a brick, it is best to cut the angle of a full brick then cut the brick to the correct length. Do not try to cut the sharp angle out of the frog side as this will normally crack and break off.

STEP 9 Build the raking cut to the wall using the line as a guide. Make sure that you fill the joints on the toothed ends properly.

Figure 5.95 Bedding on

Figure 5.96 The finished wall

Forming openings

There is usually a requirement to have either window or door openings in long walls. The bond needs to be able to reflect these openings and is generally set out dry at the setting out stage to avoid poor bonding. Any cuts should be under door or window openings. This means there should be no real difference in the bonding of the brickwork.

It is possible to cut bricks to create the dimensions needed for the **reveals**.

Problems occur if the openings are not the same size (or multiples of the size) as the bricks being used. In these cases adjustments have to be made. Broken bonds are used in short lengths of brickwork. Broken bonds are where the normal pattern of the brickwork is adjusted in order to cope with the opening. It is possible to avoid these by using reverse bonds and using dry setting out.

KEY TERMS

Reveals

– these are the sides of windows and door openings.

REED TIP

Acquiring learning skills is a lifelong experience; we're all still learning, and even after your apprenticeship is finished, you'll keep on learning too.

DID YOU KNOW?

A segmental arch is also known as a soldier arch and can be constructed using either the rough ring or axed arch methods.

REED TIP

If you want to be a good apprentice, a keen and polite attitude is essential. Listen to what people are saying, keep your eyes open, nod, smile, take it all in… and keep your phone in your pocket, or in a locker, and switched off!

The **reverse bond** allows you to use brickwork at either side of the openings when they don't match.

Window sills

Sills are usually longer than the width of the window frame. The frames are secured by using pads, slips or metal clamps.

When building a brickwork wall the opening is given a dummy frame or profile, which is made up of sawn softwood. The frame is usually larger than is actually required in order to make fitting the frame easier.

The sill is there to encourage rainwater away from the window frame and the wall below it. A cavity closer block is placed below the inner window board. This gives a large and solid base to fit the window sill.

Arches

A more ornamental way of bridging openings is to use arches, which are small units bonded together around a curve or series of curves. Arches need no additional reinforcement like brick lintels do, because the units are wedge-shaped. The more load placed on an arch, the tighter the units in the arch will become.

Arches are generally classified into three main groups according to their use or the method of cutting and preparing the arch before it is built in position:

1. Rough ring arches: the joints, and not the bricks, are usually wedge-shaped. These can be used on work which does not require a high standard of finish, or work which is to be plastered over. Little or no cutting is needed.

2. Axed arches: these are carefully set out. It uses wedge-shaped bricks (known as voussoirs), which are all cut to the same shape and size and have a pleasing appearance when finished. A semi-circular arch is constructed using this method, as seen in Fig 5.100.

3. Gauged arches: these are ornamental and expensive, as the bricks require a lot of preparatory work before they can be built in the arch, so are only used in buildings of better quality. They are prepared and bedded with a fine white joint. A sound knowledge of geometry is required before attempting this class of work.

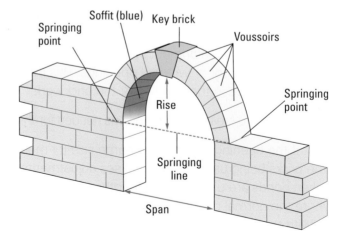

Figure 5.97 Parts of a semi-circular arch constructed using the axed arch method

Here are a few of the more common terms used in the construction of arches:

* **voussoirs**: individual straight or wedge-shaped bricks in an arch

* **span**: distance between reveals (sides) of opening that arch bridges

* **soffit**: surface of arch's underside

* **springing points**: lowest points from which arch curve starts

* **springing line**: imaginary horizontal line through springing points

* **rise**: vertical distance between springing line and highest point of soffit (under arch)

* **key brick**: highest or central brick, usually last to be built

* **crown**: highest point of arch, where key brick is placed

* **intrados**: underside edge of arch when seen in elevation – voussoirs for rough arches are usually set out on intrados

* **extrados**: upper edge of arch when seen in elevation – voussoirs for fine axed and gauged arches are set out on extrados

* **haunch**: lower part of arch measured from springing line to halfway to crown

* **bed joints**: joints between voussoirs

* **face joints**: cross joints in bonded arches, between voussoirs in arch courses

* **template**: model marked out from full-size drawing of arch and cut to shape of arch voussoirs – arch bricks are then cut to match shape of template.

When constructing a rough ringed arch, the turning piece is formed into the actual shape of the arch from a solid piece of wood. This supports the arch while it is being constructed. Turning pieces are generally used for segmental arches which have a span and rise that is less than a semicircle. It is more economical to build up a centre from small pieces of timber when you need to form arches greater than 1 m span and 75 or 100 mm rise.

The arch centre is used for the same purpose as a turning piece – as a temporary support to carry the arch during its construction over an opening. It is made up of a number of small-section timber members into the shape required. Arch centres have the advantage of being convenient to use over quite large spans.

Refer to Chapter 6 for practical exercises about building arches. The technique, using an arch centre, is much the same for solid walls and cavity walls.

Figure 5.98 This segmental rough ring arch used uncut bricks with wedge-shaped mortar joints

TEST YOURSELF

1. In traditionally built dwellings what was the approximate thickness of a single solid wall?

   a. 1,500 mm

   b. 1,000 mm

   c. 500 mm

   d. 250 mm

2. What is the main purpose of a work method statement?

   a. To comply with health and safety legislation

   b. To show the client how the job will be completed

   c. To help work out the cost of a job

   d. To provide proof that a job has been completed

3. Some brick hammers have slotted ends. What is supposed to be inserted into these ends?

   a. A spirit level

   b. A sharp cutting blade

   c. A blade or comb

   d. Another hammer

4. What tool is used to make round, sunken masonry joints?

   a. A brick tong

   b. Line and pins

   c. Bolster

   d. Jointing iron

5. If a brick's end face is exposed in a section of walling, what is the brick usually called?

   a. Frog

   b. Arris

   c. Header

   d. Stretcher

6. If you add mortar to a brick on any face, which of the following coordinating sizes is incorrect?

   a. 225 mm

   b. 112.5 mm

   c. 75 mm

   d. They are all correct

7. What is the term used to describe a brick that has been cut?

   a. Batt

   b. Transverse

   c. Shard

   d. Waste

8. What is a movement joint?

   a. A temporary joint

   b. A continuous joint filled with a compressible material

   c. A joint that bends backwards and forwards like a hinge

   d. A joint that is built in after the work is completed

9. What process involves waterproofing horizontally and vertically and usually takes place in basements?

   a. Banking

   b. Dampening

   c. Tanking

   d. Franking

10. What is often fixed to the tops of walls to make them more decorative and waterproof?

    a. Coping

    b. Coving

    c. Cementing

    d. Timber cladding

Chapter 6

Unit CSA–L2Occ731
CONSTRUCT CAVITY WALLING FORMING MASONRY STRUCTURES

LEARNING OUTCOMES

LO1/2: Know how to and be able to prepare for constructing cavity walling forming masonry structures

LO3/4: Know how to and be able to construct cavity walling to the given specification

LO5/6: Know how to and be able to form openings in cavity walling to the given specification

INTRODUCTION

The aims of this chapter are to:

* help you to select materials, components, tools and equipment
* help you to build cavity wall structures.

PREPARING TO BUILD CAVITY WALLING FORMING MASONRY STRUCTURES

Cavity walls, as their name suggests, consist of two separate walls with a space between them. The outer wall is usually brick and the inner wall is block. The walls are fixed together using metal ties.

It is the cavity itself that creates a barrier. It protects the inside of the building from water and poor weather conditions. The water cannot pass into the inner wall because the air that is circulating inside the cavity dries out any damp.

Damp-proof courses are needed wherever the walls are physically joined. This is particularly true around openings.

In practice, the cavities are not really empty. They are insulated in order to ensure that the inside of the structure remains warm and to improve its energy efficiency. To conform to Part L of the Building Regulations, current practice is that 100mm thick insulation is utilised. This will fill the cavity entirely.

Hazards, health and safety and risk assessment

Many of the potential hazards in building cavity walling are the same as those that apply to the building of other structures. There is specific information about working at height later in this chapter. You should refer to Chapter 1, pages 2–13, which dealt with health and safety.

Drawings, specifications and schedules

Once again, drawings, specifications and schedules will be your main reference material for any work you are carrying out to build cavity walling. More information about drawings, specifications and schedules can be found in Chapter 2, pages 41–7.

PPE

The exact PPE that you will need when you are building cavity walling will depend on the nature of the work. To remind yourself of the types of PPE you might need, refer to Chapter 1, pages 31–3.

Resources, tools and equipment

The range of suitable tools and equipment can be found in Chapter 5, pages 136–7, about building solid walling.

The main materials you will be using are:

* bricks
* blocks
* lintels
* cavity frames
* cavity trays
* wall ties
* damp-proof course.

We need to look at these in a little more detail.

Bricks and blocks

The type of brick or block will depend on the specific job. The bricks and blocks used for cavity walling are relatively resistant to moisture.

Other characteristics could determine whether or not they are chosen for a job. For example, they may be fire resistant or good insulators.

There are two main types of block that are used – concrete or lightweight insulation. The concrete blocks can be solid or hollow, and have different kN (kiloNewton ratings, indicating the force of gravity acting on the blocks) ratings to indicate their different strengths. Solid blocks tend to be used for commercial and industrial purposes. Hollow blocks often have rods running through them and they are then filled with concrete. Lightweight blocks have become commonplace for health and safety purposes. They also have better insulation values.

Table 6.1 outlines the main types of brick and block that you are likely to use for cavity walling.

> **PRACTICAL TIP**
>
> Some bricks and blocks are chosen because they are good at absorbing the strength of the sun. This means that they increase in temperature if they are exposed to the sun. This is known as solar gain.

Type	Characteristics
Common bricks	These are of low quality compared to other types of brick. They are used for work that is not going to be seen. They do not have a great ability to withstand heavy pressures or weights.
Facing bricks	These are the bricks that will be visible on the outside of the building. As a result they have a better and more attractive surface, which could be textured.
Engineering bricks	These are rated as either A or B. A is the strongest type of brick and B class bricks are sometimes called semi-engineering bricks. Both are ideal for damp-proof courses or for below-ground-level building work. They have a high strength, which is known as compressive strength, so they can withstand heavy pressures and weights. They are also less likely to absorb water.
Concrete bricks	They are good thermal insulators and also provide insulation from external noises. They are also fairly fire resistant.
Sand and lime bricks	These can be used instead of facing bricks and are available in a variety of colours and textures.
Dense concrete blocks	These are solid, heavy concrete blocks. They are designed to be laid up to the damp-proof course. They are also used to build the outer leaf of a cavity wall and then rendered.
Lightweight insulation block	These are less dense but good insulators and are used for the majority of the inner leaf of a cavity wall.

Table 6.1 Different types of brick and block for cavity walling

Lintels

Lintels are placed above openings in the walls. These are designed to support any brick or blockwork above the opening. They are usually made from concrete with steel reinforcement or from galvanised steel.

Cavity frames

It is not possible to fit doors or windows into walling without first putting in some linings and frames. These are designed to match the specified size of window or door and allow the component that will fill the opening to fit flush and gap-free.

Cavity trays

Cavity trays are moisture barriers. They have a lip or interlocking edge on each end. At the back of the cavity tray there is a flap shape, which can be adjusted to the inner wall. You can also install a tray using a plastic damp-proof course which comes on a roll. The roll must be wide enough to be built in to a course of blocks above the internal lintel and stepped down over the cavity to a course above the external lintel. Weep vents must be installed every 450 mm to allow moisture to drain out from the tray and let fresh air into the cavity.

The idea is that the cavity tray will encourage moisture away from the inner wall. The curve on the back achieves this.

Wall ties

There is more information on wall ties later in this chapter. But it is important to stress here that wall ties are vital when building cavity walls. Cavity walls should not be built without ensuring that both the inner and outer leaves are held together. The thickness of the wall involved means the wall would be unstable without wall ties. Wall ties are made from rust-proof metal and have twists in them. It is the twist that stops water from travelling from the outer leaf to the inner leaf. In effect this is a drip system.

Damp-proof course

Some specific advice is included later in this chapter, but it is worth stressing here that damp-proof courses are an important way of stopping moisture from entering a building. Moisture can come up from the ground, so a damp-proof course just above ground level is vital to prevent this.

The cavity between the inner and outer leaves, along with other components such as cavity trays, prevents water from getting through the walls.

Preparing and cutting components

Cutting can be done either by hand or with a machine. The exact method will depend on the block or brick you are using. Obviously you would not try to hand saw a dense concrete block or engineering brick. For these you would have to use a bolster chisel and hammer. Other blocks and bricks can be cut using disc cutters or table saws.

Angle grinders are also very useful and are used mainly for cutting bricks and other blocks. Some construction sites have petrol cutters. These have extra-strong diamond cutting discs.

Protecting the work and surrounding area from damage

It is important to remember that when carrying out your work you should always try to bear in mind these points:

* Minimise damage to the area – only clear ground right next to you to give you enough working space.

* Maintain a clean working space – do not leave the working space cluttered with debris, tools, materials and equipment. Always tidy up after yourself to prevent contamination of the site and health and safety hazards.

* Dispose of waste – sort your waste into types. Some of it, such as plastic, can be recycled. Bits of brick and block may be useful for hard core. You should have separate bins or skips for particular types of waste on site.

* Make the construction sustainable – by following correct building procedures and ensuring your work is of top quality. The building should have a long and trouble-free life. It will also be energy efficient and meet all environmental challenges.

BUILDING CAVITY WALLING TO THE GIVEN SPECIFICATION

Walls may be built as solid walls, that is, having one thickness only, or cavity walls, being made of two walls (not necessarily of equal thickness), with a space in between. These walls are called leaves and are generally referred to as the outer leaf and inner leaf. The space or hollow in between the walls has two main advantages:

1. When the outer leaf becomes wet, the space prevents water passing from the outer wall to the inner wall, and so the inside of the building remains dry.

2. The space will not allow heat to be transferred from one wall to the other. Air is a bad conductor of heat and the cavity acts as a barrier. The cavity, therefore, is a good heat insulator and helps to keep the building cool in summer and warm in winter.

If the cavity is sealed at the base, eaves, and all the jambs of openings, then the air trapped inside will be static and this will ensure good heat insulation for the building.

The cavity should be large enough to provide suitable insulation, but not so wide as to make it difficult to tie the two walls together. The Building Regulations state that the width of the cavity at any level should be not less than 50 mm or more than 100 mm where ties are placed at distances apart not exceeding 900 mm and 450 mm vertically.

A cavity may be not more than 100 mm wide if vertical twist-type ties are used and are placed at distances apart not exceeding 750 mm measured horizontally and 450 mm measured vertically.

Foundations

Cavity walls for housing may be built directly off the foundation concrete; no footing courses are considered necessary, as long as the ground immediately below the foundations is suitable for carrying the load of the building. In these cases the cavity is usually filled with a concrete consisting of a small aggregate, not exceeding 10 mm, and cement. One part cement to six parts aggregate is suitable.

The filling is taken up to ground level and finished off to a steep slope, falling towards the outer leaf at the top. This slope will direct any accumulation of water which may collect inside the cavity away from the inner leaf. Water may enter the cavity when rain passes through the outer leaf, runs on to the wall ties, and drips off the middle of the ties into the cavity.

Condensation

Moisture may also develop from condensation, when the warm air in the cavity is cooled on the inside face of the outer leaf of the cavity wall. Warm air holds more water than cold air, so when the temperature of the air is lowered, water is released in the form of droplets, known as condensation.

It is important to dispose of this water as soon as possible and prevent it collecting in the base of the cavity, so, in addition to the slope on the top of the filling, weep holes are sometimes left in the outer skin of the wall at the first course above the slope of the filling. The weep holes are formed by raking out the mortar in the perpend joints about every metre along the wall. The joints should be completely free from mortar so that the water can drain away quite freely.

An alternative method is to build the wall solidly at the base and then begin the cavity at ground level, or at 150 mm below the damp-proof course, as shown in Fig 6.1.

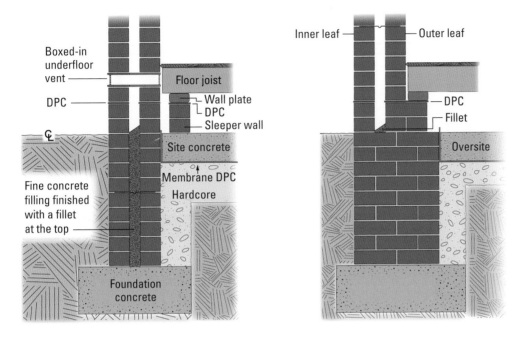

Figure 6.1 Sections showing alternative methods of stabilising the base of a cavity wall

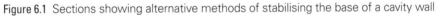

Building cavity walling

Blockwork should be laid out at half bond. It is usual practice to lay out the blocks before mortaring them into place, in order to look at the bond possibilities. Usually a cut piece of block will be needed to complete the course.

Straight lengths
When you finish with a full block on the course below, you build the wall with straight joints, using a half block. This means that you work back towards the first corner.

In Fig 6.2 load-bearing insulation blocks are used for the internal part of the cavity wall, (marked A). Foundation blocks 250 mm thick or greater are used for cavity walls (marked B). Load bearing foundation blocks are required below load bearing walls (marked C). The load-bearing walls need blocks of between 100 and 250 mm (marked D). Standard blocks are used for non-load bearing walls and can be between 75 and 250 mm (marked E). Aerated blocks can also be used.

See page 235 for information about returns and junctions.

The following practical tasks show you the steps you need to take in order to build a variety of different types of cavity walling. Always check regularly that the walling is both level and plumb.

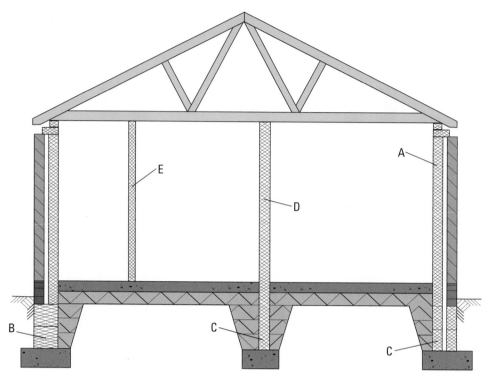

Figure 6.2 Section view of a house showing blockwork

1. BUILD A CAVITY WALL WITH PARTY WALL JUNCTION

OBJECTIVE

To build a straight section of cavity wall. The block leaf is to have a party wall junction installed.

This can be seen on buildings where adjoining properties require a cavity wall that is filled with insulation for both sound and insulation properties.

Ensure you select PPE appropriate to the job and site where you are working. Refer to the PPE section in Chapter 1.

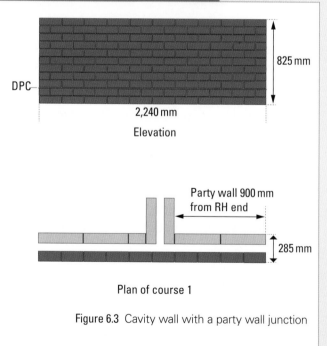

Figure 6.3 Cavity wall with a party wall junction

TOOLS AND EQUIPMENT

Walling trowel	Spirit level	Straight edge
Lump hammer	Corner block and lines	Jointing iron
Bolster chisel	Builder's square	Steel tape measure

STEP 1 Set out and dry bond the wall 2,240 mm long. Bed the first brick, level and to gauge at the highest end of the floor. Then bed another brick at the other end. Level the second brick with the first using a spirit level and straight edge.

Check the alignment with the straight edge and then re-check all the previous criteria so that length, level and alignment are all correct before walling starts.

STEP 2 If you just put the corner blocks and line onto these two bricks when the mortar is still wet, it will pull them out of place. You can overcome this by placing two or three bricks onto the top of the two end bricks to stop them from moving.

The bricks can then be laid in to complete the first course.

STEP 3 Build the second course to DPC height (150 mm).

STEP 4 Repeat **Steps 2** and **3** for the inner leaf.

The inner leaf of the cavity must also be two courses of bricks high so that the subsequent brick and block courses are correct.

Set out the party wall 900 mm from the right hand end of the model and checked for square to the main wall.

PRACTICAL TIP

If you use the spirit level and straight edge to transfer the level, remember to rotate it 180° in order to eliminate any possible inaccuracies.

STEP 5 Lay a thin screed of mortar on top of the second course so that the DPC can be laid. As you are laying the DPC, gently run your walling trowel over the top of it and press the DPC into the mortar screed.

This is not to stick the DPC down, but to protect it from high spots or sharp areas that may puncture it and stop it working. Any overlaps in the DPC should be a minimum of 100 mm.

Figure 6.4 Laying the DPC

STEP 6 Build the block corners first. When running the blockwork in, the party wall section should be left until last on each course and then checked to the line with a spirit level. If you lay the blocks at the party wall then the line could snag on the blocks and the rest of the blockwork would be inaccurate.

STEP 7 Wall in the blocks between the corners using corner blocks and line.

Place wall ties 225mm vertically at the unbounded ends of the cavity, and at 450mm vertically (every two course of blocks) throughout the body of the wall. Put in a diamond pattern as much as possible, so that the ties are taking equal loads. Horizontal spacing should be 900mm max and the ties at the stop ends should be placed 225mm from the edge, so as not to interfere with any methods of closing the cavity.

STEP 8 Now the brickwork can be constructed. As usual, build small corners first and then run in with the line in order to gain the best efficiency.

Figure 6.5 Placing block at junction

Figure 6.6 Levelling the blocks

STEP 9 The wall should be finished with a half-round joint on the brickwork and a flush finish on the block leaf. Brush the wall when the mortar is dry enough to do so without leaving any brush marks.

PRACTICAL TASK

2. BUILD A CAVITY WALL WITH EXTERNAL RETURN

OBJECTIVE

To complete a section of cavity wall to the required standards. The wall will include an external return as well as a junction.

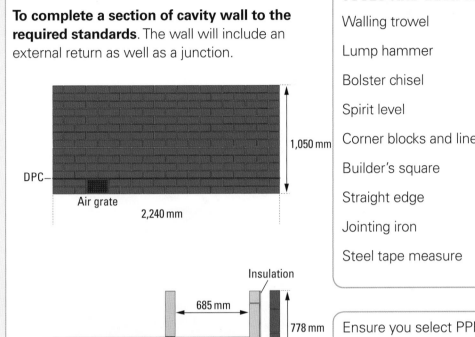

TOOLS AND EQUIPMENT

Walling trowel

Lump hammer

Bolster chisel

Spirit level

Corner blocks and line

Builder's square

Straight edge

Jointing iron

Steel tape measure

Ensure you select PPE appropriate to the job and site where you are working. Refer to the PPE section in Chapter 1.

Figure 6.7 Cavity wall with external return

The measurements required are:

Length of wall 2,240 mm

Height of wall 1,050 mm

Width of cavity overall 300 mm

Lengths of brick return 778 mm

STEP 1 Set out and dry bond the wall 2,240 mm long. Bed the first brick, level and to gauge at the highest end of the floor. Then bed another brick at the other end. Level the second brick with the first using a spirit level and straight edge.

Check the alignment with the straight edge and then re-check all the previous criteria so that length, level and alignment are all correct before walling commences.

STEP 2 Lay the bricks to complete the first course. The return corner can then be established to the correct length and checked for being 90° with either a builder's square or the 3:4:5 method. The air brick should be dry bonded into position and can then be run in with the second course.

Build the second course to DPC level (150 mm).

Figure 6.8 Air brick

STEP 5 Lay a thin screed of mortar on top of the second course so that the DPC can be laid. As you are laying the DPC, gently run your walling trowel over the top of it and press the DPC into the mortar screed

This is not to stick the DPC down, but to protect it from high spots or sharp areas that may puncture it and stop it working. Any overlaps in the DPC should be a minimum of 100mm.

A damp tray should be cut over the air brick, so that the ends can be turned up into the cross joints, and the upper section bedded into the block course in the inner leaf.

STEP 4 Repeat **Steps 2** and **3** for the inner leaf. The inner leaf of the cavity must also be two courses of bricks high so that the subsequent brick and block courses are correct.

The liner for the air brick should then be positioned opposite the air brick.

Figure 6.10 Fitting the damp tray

Figure 6.9 Air brick liner

STEP 6 Build the block corners first so that the insulation can be applied before the outer leaf is formed.

When running the blockwork in, leave the return on each course until last and then check it to the line with a spirit level. This prevents the accuracy of the line being compromised and leading to any error.

PRACTICAL TIP

It may help to position the wall ties on top of the DPC to support the insulation. On site, the insulation may protrude below DPC and the tie will be applied accordingly.

STEP 7 Place wall ties 225 mm vertically at either of the unbonded ends of the cavity and the ties at the stop ends should be placed 225 mm from the end, so as not to interfere with any methods of closing the cavity, and at 450 mm vertically throughout the body of the wall. Put them in a diamond pattern as much as possible, so that the ties are taking equal loads. Horizontal spacing should be 900 mm max but may be reduced to support the insulation.

STEP 8 Apply the insulation and cut it to fit so that it is bonded in much the same way as stretcher bond is. It should be cut at the ends so that you can fit any cavity closers without having to alter the insulation. The amount to be cut depends on the depth that the closer sits in the cavity, but 50 mm should be sufficient. Then the insulation can be fixed with the appropriate plastic clips.

STEP 9 Construct the brickwork. As usual, start by building small corners and then run in with the line in order to gain the best efficiency.

You will have to insert the cavity closers at the same time as you build the corners. Most closers come with plastic clips that need to be built into the bed joints of the brickwork.

STEP 10 Finish the wall with a half-round joint on the brickwork and a flush finish on the block leaf. Brush the wall when the mortar is dry enough to do so without leaving any brush marks.

PRACTICAL TASK

3. BUILD A CAVITY WALL WITH RETURN AND RUSTICATED CORNER

OBJECTIVE

To build a cavity wall with a return corner, with the addition of a rusticated corner and a window with a soldier course.

Both the quoin and soldier course are mainly decorative features that can enhance the appearance of what can be rather plain cavity walls.

TOOLS AND EQUIPMENT

Walling trowel

Lump hammer

Bolster chisel

Spirit level

Builder's square

Corner blocks/pins and line

Straight edge

Jointing iron

Steel tape measure

Boat level

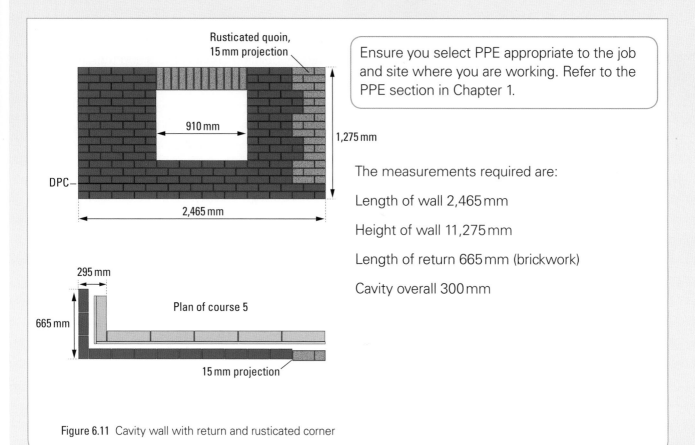

Figure 6.11 Cavity wall with return and rusticated corner

In the figure:
- Rusticated quoin, 15 mm projection
- 910 mm
- 1,275 mm
- DPC
- 2,465 mm
- 295 mm
- 665 mm
- Plan of course 5
- 15 mm projection

Ensure you select PPE appropriate to the job and site where you are working. Refer to the PPE section in Chapter 1.

The measurements required are:

Length of wall 2,465 mm

Height of wall 11,275 mm

Length of return 665 mm (brickwork)

Cavity overall 300 mm

STEP 1 Follow **Steps 1** to **6** from the practical task, Build a cavity wall with external return, on page 187.

STEP 2 Lay the first course of blocks placing the wall ties where necessary.

Mark out the position of the opening on the finished course of blocks with the aid of a thin mortar screed.

PRACTICAL TIP

It's good practice to make the opening on the blockwork slightly bigger so that it does not project past the brick opening on the outer skin.

STEP 3 Complete the blockwork.

Leave half a block of space at either side of the opening for the lintel bearing at the top of the block wall.

STEP 4 Install the insulation and fix it with the appropriate ties.

STEP 5 Set out the corner using a contrasting brick for the rustication with a projection of 15 mm. Build a small return corner.

At this stage it is better to build up the three courses on the rusticated corner and complete the brickwork in between them.

It is essential that the inside edge of the projection is also plumbed so that the corner has a plumb line down the block-bonded quoin.

STEP 6 Bed the lintel on a thin screed of mortar, so as not to make the bed joint of the soldier course too high.

If the lintel is rocking, place some mortar under the opposing corner to make it stable.

STEP 7 Build the last three courses of bricks. Make sure that you finish at the sides of the opening and that you make sure the ends are plumb.

Place the damp-proof tray across the lintel with the ends turned up into the cross joints.

Figure 6.12 Building cavity tray over lintel

Figure 6.13 Placing weep holes

STEP 8

Check the width of the opening with a tape measure to make sure that the soldier course will fit. You may need to make the joints larger or smaller to lay the soldiers fit with evenly spaced joints. Ideally the opening should be 910 mm for the bricks to fit correctly with a 10 mm joint.

Figure 6.14 Checking length of the soldier course

Figure 6.15 Checking level of the soldier course

Figure 6.16 Laying the soldier course with weep holes

STEP 9 Lay the soldier course. Some bricklayers will use two lines for walling in the soldier course, one at the top and one about two-thirds of the way up from the bottom.

Keep checking each brick with a boat level for plumb.

Add weep holes after the first soldier from each end and place one in the middle of the opening.

STEP 10 Check the bottom alignment with a suitable straight edge such as a spirit level.

STEP 11 Finish the brickwork with a half-round joint to the brickwork and a flush joint to the blockwork.

PRACTICAL TIP

Always select the bricks for the soldier course. They should be the same size and have good straight edges on the face of the brick. This will make it easier to lay the soldier course.

It is good practice to stand back every three bricks and take a look at the bricks to make sure that they look right.

When laying the soldiers, it helps to get the cross joint as close to perfect as possible, before lowering it to the line. Angling the mortar, so that there is more mortar at the front of the bed, will help to prevent bricks from tipping forwards.

PRACTICAL TASK

4. BUILD A CAVITY WALL WITH A GABLE END

OBJECTIVE

To develop practical skills by building a cavity wall with a gable end.

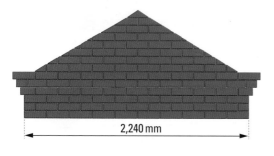

2,240 mm

300 mm

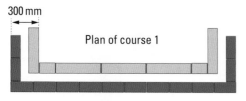

Plan of course 1

Figure 6.17 Cavity wall with a gable end

TOOLS AND EQUIPMENT

Walling trowel	Straight edge
Lump hammer	Jointing iron
Bolster chisel	Steel tape measure
Spirit level	Builder's square
Corner block/pins and line	Appropriate sized roof truss

Ensure you select PPE appropriate to the job and site where you are working. Refer to the PPE section in Chapter 1.

The measurements required are:

Length of wall 2,240 mm or to suit the temporary roof truss

Width of cavity 300 mm

Overhang for the oversail 36 mm

STEP 1 Position the timber roof truss that is representing the gable end of a house, then mark the position of the inner wall in relation to it.

Figure 6.18 Roof truss

Figure 6.19 Positioning the temporary roof truss

STEP 2 Bed the blocks at either end and run in to form the first two courses.

PRACTICAL TIP

If there are no 10 mm packers available, the block or brick can be positioned on the course underneath it with a 20 mm spacing from the adjacent block/brick. When cut and positioned, this will allow for a 10 mm bed and cross joint.

It is not necessary at this stage to build the block skin to full height if the wall is becoming unstable. On site, large gables would also be built in sections in this way and they would be anchored to the roof trusses with galvanised steel fixing brackets.

STEP 3 Build the blocks in front of the truss, cutting the end blocks to the angle of the truss. When bedding the blocks, make sure that they do not protrude above the roof truss or the roof tiles will not fit onto the roof.

When cutting the blocks place the blocks onto 10 mm packers of timber to maintain accuracy and mark the block in line with the angle of the roof truss.

Cut the blocks, saving any off-cuts that could be used on another section and lay to the line.

Figure 6.20 Positioning the blocks

STEP 4 Set out the bricks with the correct spacing to form the cavity required and lay the first course.

STEP 5 Build the corners for the first three courses and run in the brickwork between the corners.

STEP 6 Build the oversail courses with an overhang of 36 mm on each course and run in the brickwork between the corners.

Figure 6.21 Building the oversail

STEP 7 As with the blockwork, build in the full bricks to the gable first. On a large gable, small pyramid corners can be built and run in to the line.

STEP 8 Fix a timber strut at the top and bottom of the roof truss. A line can be attached for marking the bricks to the raking cut.

Figure 6.22 String a line to the gable

STEP 9 Mark the brick to be cut using the spacers or the 20 mm method and cut to suit, while saving any off-cuts in case they can be used on another section. Bed the cuts to the line, checking that they are level and are not above the line of the roof truss.

PRACTICAL TIP

To check for any high spots on a cut, eye the back arris of the cut with the front while closing one eye. This will enable you to spot any extra material that will need to be removed from the cut face.

STEP 10 Build to the top (along with the block work if the gable has been built in stages) and finish with a half-round joint, brushing it when ready.

Figure 6.23 Building up the courses to the gable

Figure 6.24 The finished wall

Figure 6.25 Insulation in the cavity

CASE STUDY

Tips for getting your cavity walls right

Marcus Chadwick is a bricklayer at Laing O'Rourke.

'You must make sure your setting out is right, make sure your DPC (damp-proof course) is in place. Make sure your positioning of the ties is correct from outer leaf to inner leaf, that your insulation is pegged back properly, and you've put discs over your ties.

Always make sure your cavities are clear – never leave what we call 'snots' on the back of the brickwork. If any more drop down on top of them, it's a collective process – it could bridge the cavity and that's when you'll start getting damp penetration to the inner leaf and ultimately damage the plaster. Cleanliness of the cavities is paramount.

In order to go back down to ground level and clean it out, we leave the odd brick out on top of the DPC so we can do a cavity clean once we've got the wall to height. We get some hessian rolls, cut a length out, feed it through from the outer leaf through the missing brick, into the cavity and then back out. We keep doing that so it forms a continuous loop, and then when we've finished the job, we pull the hessian out and it brings all the waste mortar with it, so there's no chance of any bridging. And it keeps your boss happy too!'

Damp-proof barriers

Wall ties need to be free from mortar droppings when the wall has been completed. If this is not checked then the mortar can form a bridge across the cavity. This bridge will allow water to pass across the cavity and the inner leaf may become damp.

Other places where dampness may penetrate to the inner leaf are at reveals, heads of openings and sills. Therefore a damp-proof membrane should be provided at these points so that the interior will be kept as dry as possible. Figs 6.26 to 6.30 show methods of sealing the cavity at these places.

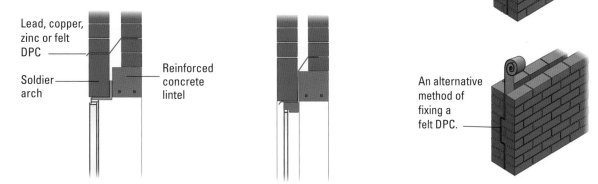

Figure 6.26 Treatment at window heads to prevent dampness penetration

Figure 6.27 Methods of sealing cavities

To give added stability to the two leaves it is also advisable to seal the cavity at roof level by bridging across the two leaves. However, you need to do this above the level of the eaves so that it is then not necessary to insert a damp-proof course where the cavity is bridged at this level (Fig 6.30). However it may be that the drawings require the cavity and roof insulation to be linked; in this case, the cavity cannot be sealed at the top.

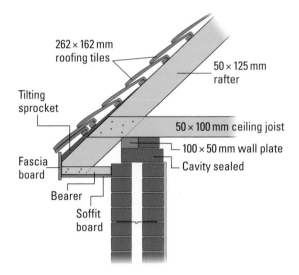

Figure 6.30 Method of sealing a cavity at eaves level

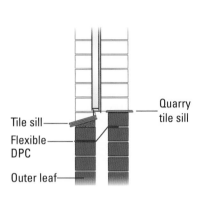

Figure 6.28 Treatment at window sill to prevent dampness penetration

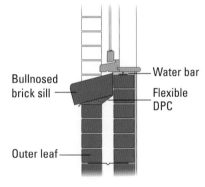

Figure 6.29 Section showing typical treatment at a brick sill to prevent the penetration of water

Insulation

Under Building Regulations it is a requirement to ensure that cavity walls are insulated. The Building Regulations outline, according to specification, exactly how much insulation is needed.

During the construction of blockwork walling, wall ties are used to stabilise the structure. A total fill requires the cavity to be filled with insulation slabs or boards, with the wall ties positioned above each section of insulation. Insulation slabs or boards come as standard in 450 mm × 1,200 mm sizes and are made from mineral fibres.

An alternative is to partially fill the cavity walling. Wall ties are used to keep the insulation in place but the insulation slab or board is held against the inner wall by plastic clips.

The final alternative is to inject insulation into the cavity after construction. This is usually polystyrene granules or rock wool fibreglass. Holes are drilled into the inner walls and the insulation is pumped into the cavity. This is the technique used for new builds.

In order to make sure that the walling is suitable, the cavity has to be at least 50 mm wide.

The masonry and brickwork need to be in good condition for insulation to be added.

When insulation is fitted to existing buildings small holes (approximately 22 mm in diameter) are drilled into the walls at intervals of around 1 m. Insulation is then blown into the cavity and the holes are filled.

Typically, insulation for cavity walls can be made from three different types of material:

* mineral wool

* beads or granules

* insulation foam.

When insulation is being fitted into new cavity walling it needs to comply with British Standards and will usually have a guarantee of 25 years.

Decorative features

There are many ways in which decorative features can be incorporated into a cavity wall construction project. These are often called design detailing. They aim to give distinctive features to the external appearance of the building. Table 6.2 outlines some of the ways in which this can be done.

Decorative feature	Description
Dog tooth courses	These are courses of bricks that consist of cut bricks laid at 45° to the main face of the wall. This makes the bricks look as if they are triangular in shape.
Quoins	These are bricks that project from the face of the external wall. They can be in a contrasting colour.
Polychromatic brickwork	These are bricks of different colours, so that patterns on the face of the wall can be created. The bricks can also be bedded in with different coloured mortars to match or contrast with the bricks.
String courses	These are bricks that project slightly beyond the face of the main wall. They can be different colours, shapes or even made with panels, and some are textured.
Soldier courses	This is a row of bricks orientated in the same direction and placed on their sides. They are commonly used for edging.
Dentil courses	These are projecting alternate headers put in to form regular patterns.
Band courses	These are decorative bricks that are flush with the main wall. They are usually a slightly different length from other bricks in the wall.

Table 6.2 Decorative features for cavity walling

Figure 6.31 Dog tooth courses

Figure 6.32 Quoin courses

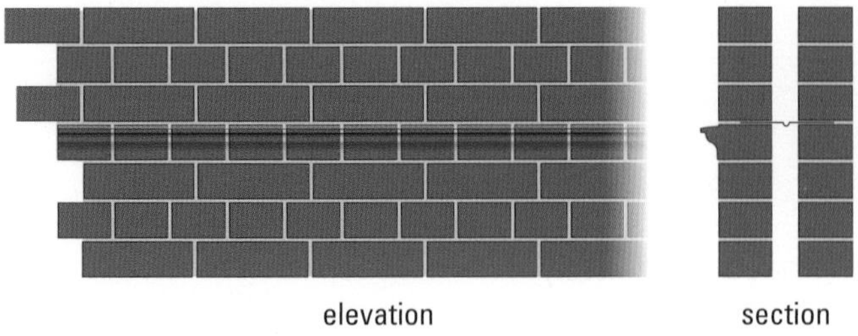

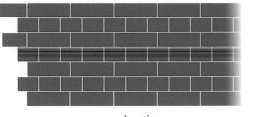

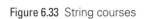

elevation
section

Figure 6.33 String courses

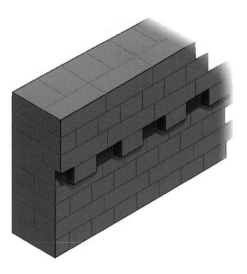

Figure 6.34 Dentil courses

REED TIP

Make sure you look after your tools or your tools won't look after you.

It is important to remember that with these decorative features the bricks are always cut flush with the inner face of the outer leaf. This is to ensure that the cavity is not bridged.

Wall ties

Wall ties are used to tie the walls together, or to stabilise them. They are generally made of a non-ferrous metal or stainless steel. The illustration in Fig 6.35 shows various types of ties that are used in cavity walling; all types have one thing in common, in that they are designed to trap the passage of water from the outer to the inner leaf (see Fig 6.36 on page 199). This is an essential requirement if the cavity is to function properly and keep the inside wall dry.

Simple galvanised wire wall tie
Catnic stainless steel wall tie
150 or 200 mm
75 or 100 mm
Butterfly or galvanised wire wall tie
Twist in the middle of the wall tie
Galvanised iron tie
21 mm
Double wall triangle tie
150 or 200 mm
37 or 50 mm
Vertical twist wall tie

Figure 6.35 Types of cavity wall ties

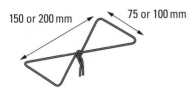

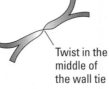

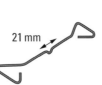

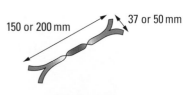

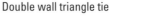

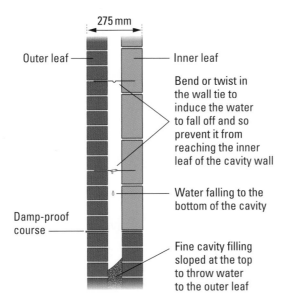

Figure 6.36 Ties trap the passage of water from the outer and inner leaf

The wall ties must be placed close enough together to ensure that stability in the wall is maintained. Building Regulations require ties to be placed at distances not exceeding 900 mm horizontally and 450 mm vertically.

In practice it is found to be an advantage if the ties can be staggered.

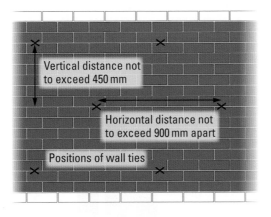

Figure 6.37 Elevation of a cavity wall showing the wall ties staggered throughout the wall

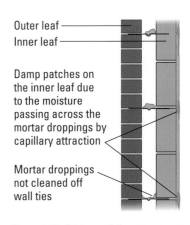

Figure 6.38 Dirty wall ties can cause damp patches in a wall

As already mentioned, when cavity walls are being built it is essential that the wall ties are kept absolutely free from mortar droppings on completion of the walling; otherwise this can cause a bridge across the cavity which may allow water to pass across and cause the inner leaf to become damp (see Fig 6.38).

You can keep cavities reasonably clean and free from mortar droppings if you continually use cavity battens while raising the work. Beginning at the lowest part of the wall, a length of batten is placed on the first level of wall ties to catch the mortar droppings as the cavity wall is built to the next level of ties. At this stage the batten is carefully raised, the loose mortar is cleaned off and returned to the mortar boards, and the batten is replaced on the new layer of wall ties. The same procedure is carried out at the level of the wall ties (see Fig 6.39). At the end of the day's work, you should insert a long batten into the cavity and tap each of the wall ties to dislodge any mortar which may have fallen on top of them.

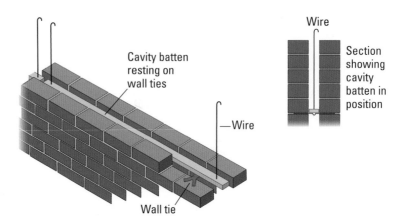

Figure 6.39 Method of keeping a cavity wall clean during construction

In a similar way, the bottom of the cavity should be kept clean. This needs to be done while the wall is built, so cleaning holes should be provided at frequent intervals in the first course of the fair work above the cavity filling; this allows the cavity to be thoroughly cleared of mortar droppings. These holes are filled in by brick inserts after the wall is completed. If the mortar droppings are allowed to rise above the damp-proof course, the rising damp may penetrate the inner leaf and also the flooring, which may be the prime cause of dry rot in a timber floor (see Fig 6.40).

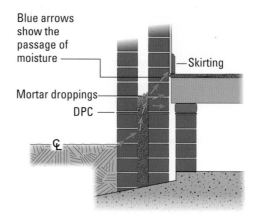

Figure 6.40 How moisture can penetrate to the inner leaf and cause damage to the timber floor

Working at height

When you are working on cavity walling it may be impossible to avoid working at height. At all times you should use equipment and safeguards and any equipment that might minimise the distance that you could fall.

Ladders can be used but should generally be your last option. Mobile elevating work platforms, which can easily be moved around, provide very safe access when carrying out high level work.

If a Health and Safety Executive inspector visits a site they will be looking to see that:

* the right risk control measures are in place, including risk assessments carried out

* anyone working at height has access to the correct equipment and that they are actually using it

* anyone working at height is competent enough to be using the equipment

* any equipment is regularly being inspected and maintained.

Above all, every employer has a duty of care to look after the health and safety of anyone while they are at work.

> **PRACTICAL TIP**
>
> Various specialist pointing tools can be used. A pointing trowel can be used to make cuts in the mortar, or a tool known as a Frenchman is used to cut straight lines.

FORMING OPENINGS IN THE CAVITY WALLING TO THE GIVEN SPECIFICATION

Setting out

Typically openings in cavity walls are created so that it is possible to fix doors and windows, or perhaps as a decorative feature in part of the wall.

The door and window openings are created during the construction of the cavity wall. Generally door and window frames can be either:

* fitted as the work on the wall is underway, so a frame is actually built into the walling

* created after the wall has been built by cutting the opening.

Forming openings in cavity walling

Frames act as profiles, which mark the opening that is required in the walling. These are plumbed shapes and squared off. They are held in position and the wall is then built around the frame.

The alternative is to fix built-in frames as the courses of brick are being built. These frames have **heads** and **sills**, which are longer than the actual width of the frame. These additional lengths are known as horns. They are trimmed and shaped to ensure that the facing bricks cover the head.

The frames are held in place using either metal cramps or wooden pads.

Metal cramps

Metal cramps are fixed onto the back of the frame. The cramp is positioned so that it is in line with a horizontal joint in the brickwork. This means that bricks are then laid on top of the cramp.

Wooden pads or slips

Pads are 90 mm square pieces of timber that are 10 mm thick. These are also placed so that they coincide with a horizontal brickwork joint. The pads are put in every four to six courses.

These pads are useful as they allow the frame to be put in later.

Bridging openings with steel and concrete lintels

Lintels support the load above an opening. Lintels can be made from a variety of different materials:

* Timber – these are more traditional and made from a solid beam.

* Stone – another more traditional material comprising of a solid block.

* Reinforced concrete – a more modern alternative.

* Steel – made from galvanised material.

The folllowing point about lintels should be noted:

* Concrete and steel can be combined and can support an enormous weight. This allows the construction of large openings.

A lintel has to be capable of supporting the load placed on it. It must not bend or deflect that load. In the past, when smaller windows were used in older houses, wood was an acceptable material for a lintel. But over time wooden lintels become weaker due to rot and to warping.

Stone lintels are also only really strong enough for relatively narrow openings. In many houses where stone lintels can be seen they are in fact supported by galvanised steel.

It is possible to use brick lintels, but these are considered to be non-load-bearing. They are must be supported by a steel lintel if they are to be load-bearing. Bricks are ideal if a shaped arch in an opening is required.

In most modern buildings concrete that has been reinforced with steel tends to be used. The steel reinforcement is positioned approximately 25 mm from the bottom of the lintel. There are many different types of concrete lintel that are designed to match particular brick or blockwork or wall thickness.

Figure 6.41 Steel lintel being built in

For larger openings pre-stressed lintels can be fitted. These have wire strands that have been set into the concrete under tension.

Galvanised steel lintels are also widely used. They are comparatively lightweight. Large openings are fitted with heavyweight rolled steel joists, more commonly known as RSJs.

Figure 6.42 Rolled steel joists (RSJs)

Brick and proprietary sills

Sills are designed to encourage rainwater to fall away from the wall below the opening. Windows and doors have sills.

Cavity walls have what is known as a closer block, which is bedded in below the inner window board. This provides a strong base for the window board. Traditionally window sills would have been wooden. Brick-on-edge sills can be fitted underneath the wooden window sill to make sure that all of the water runs away from the window frame. This also prevents the wooden window sill from rotting.

PRACTICAL TASK

5. BUILD A CAVITY WALL WITH JUNCTION ATTACHED BY FIXINGS

OBJECTIVE

To build a section of cavity wall with a junction attached with wire mesh bonded into the bed-joints on one pier and an attached fixing to the other.

This method can be used when the block type differs from those used in the cavity leaf, to reduce the risk of cracking. It also removes the risk of poorly filled joints that can occur if indents are used.

Ensure you select PPE appropriate to the job and site where you are working. Refer to the PPE section in Chapter 1.

TOOLS AND EQUIPMENT

Brick trowel	Wall starter
Lump hammer	Block/pins and line
Bolster chisel	Straight edge
Spirit level	Jointing iron
Builder's square	Steel tape measure
Mesh reinforcement	Boat level

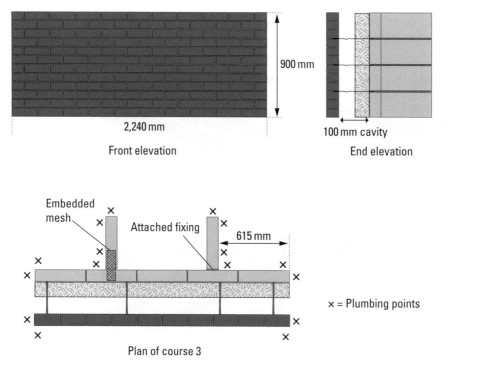

Figure 6.43 Cavity wall with fixings and mesh model

STEP 1 Follow **Steps 1** and **2** of the practical task, Build a cavity wall with party wall junction, on page 185.

STEP 2 Build the brick leaf to three courses high to make the setting out of the block leaf easier.

STEP 3 Repeat the setting out procedure with the blocks so that the block leaf is also 2,240 mm long and directly opposite the outer leaf with the correct sized cavity.

PRACTICAL TIP

Check the blocks for upright with a level on the first course as this will make plumbing of further courses easier.

STEP 4 Build up the block corners with the ties placed in the correct position. Where the block wall will incorporate the mesh for the junction, build the mesh in as the wall is constructed.

STEP 5 Run in the block walls to full height with the line.

STEP 6 Attach the insulation in the correct staggered fashion and attach with the appropriate clips. Appropriate clips are normally supplied with the wall ties or insulation, as different manufacturers have different ones.

STEP 7 Repeat on the brick leaf by building small corners and run the courses in to the line.

STEP 8 Build the mesh into the wall to take the attached wall, making sure that the mesh is fully bedded into the mortar beds, preferably to the full width of the block leaf, and checking it for square.

PRACTICAL TIP

Although the blocks can be cut to maintain bonding on alternate courses of the junction, they will not be stronger than the blocks will be if left uncut. Therefore they can be stack-bonded without affecting the strength of the pier.

STEP 9 Mark the position and then fix the wall starter with the correct fittings supplied, making sure that it is secure, plumb and runs in line with the middle of the pier.

STEP 10 Lay the first block, check for square and then build the pier with the ties correctly slotted into position.

STEP 11 Apply a half-round joint to the brickwork and finish the blockwork with a flush joint.

PRACTICAL TASK

6. BUILD A CAVITY WALL WITH A ROUGH RING SEMI-CIRCULAR ARCH

OBJECTIVE

To construct a semi-circular arch formed from brick and block cavity construction.

In a rough arch like this one, the curve is formed by the joints being shaped to form the arc rather than the bricks themselves being shaped.

Ensure you select PPE appropriate to the job and site where you are working. Refer to the PPE section in Chapter 1.

TOOLS AND EQUIPMENT

Walling trowel	Adjustable bevel
Lump hammer	Steel tape measure
Bolster chisel	Compass dividers
Spirit level	Scutch hammer
Corner block/pins and line	Timber arch former and support wedges
Builder's square	
Straight edge	
Jointing iron	

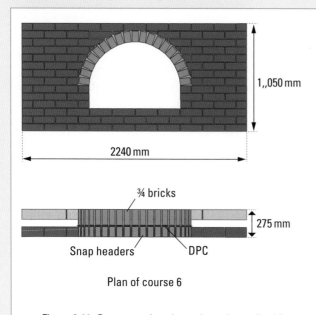

1,,050 mm

2240 mm

¾ bricks

275 mm

Snap headers   DPC

Plan of course 6

Figure 6.44 Cross section through cavity wall with a rough ring semi-circular arch

The measurements required are:

Length of wall 2,240 mm

Height of wall 1,050 mm

Width of cavity 300 mm

STEP 1 Follow **Steps 1** to **4** of the practical exercise, Build a cavity wall with party wall, junction on page 185.

STEP 2 Build three courses of bricks and one course of blocks, adding ties as required until the height has been reached for the arch opening to be added.

STEP 3 With the opening width established, add the next three courses of bricks and one course of blocks to reach the springing point of the arch.

STEP 4 Position the arch centre in the opening and adjust it to the correct height with blocks or bricks placed under it. Finally use wooden wedges to make any fine adjustment to get the centre level in both directions. Make sure that the face is plumb and does not protrude past any subsequent brickwork that is to be laid.

Figure 6.45 Using the arch centre

STEP 5 Build the brick corners up as high as possible, with allowance for the bricks that are to go around the arch.

STEP 6 Mark the centre of the arch and place a key brick over this mark. There should always be an odd number of bricks in each ring.

Use the compass dividers to gauge the spacing of the bricks over the edge of the centre and mark with a pencil when an even spacing has been obtained.

Figure 6.46 Marking the spacing around the former

STEP 7 Start adding the snapped headers (rough cut half bricks) to the marked lines and shape the joints to maintain the bricks' accuracy over the arch. Maintain alignment with the line across the face.

Figure 6.47 Building up the arch

PRACTICAL TIP

Start with small spacing on the underside and open the spaces up as required to help avoid large joints over the arch. Mark the spaces in from either end towards the centre.

Figure 6.48 Bricks meet around the outside of the arch

PRACTICAL TIP

Work from each side of the arch towards the centre before adding the cuts over the arch, rather than trying to cross the arch in one go. Make sure they are cut accurately to avoid problems when closing the cavity.

STEP 8 Cut the bricks over the arch by allowing for a constant 10 mm joint over the arch bricks. Cut roughly to shape with a hammer and bolster and then finely trim with the scutch hammer.

Figure 6.49 Marking a curve on a brick to be cut

STEP 9 Once clear of the arch, continue building the brickwork until the finished height has been reached.

STEP 10 On the blockwork side, build the corners up while allowing for the cut bricks to be added that will eventually close the cavity.

STEP 11 Cut a length of 150 mm DPC to be placed between the brick inner and the outer leaf of the arch.

Using the line to maintain alignment, start laying three-quarter bricks over the arch centre on the inner leaf. Make sure that you put a mortar joint between the brick and the DPC so as not to puncture it.

STEP 12 Fill in the cut blocks to the remainder of the block leaf. Remove the timber arch former.

Figure 6.50 The finished wall

Figure 6.51 Another example (1)

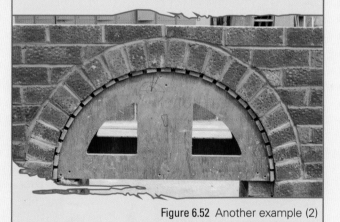

Figure 6.52 Another example (2)

PRACTICAL TASK

7. BUILD A CAVITY WALL WITH A ROUGH RING SEGMENTAL ARCH

OBJECTIVE

To construct a rough segmental arch.

This means that the bricks over the arch are not shaped and it is the forming of the joints that allows the arch to be shaped correctly.

> Ensure you select PPE appropriate to the job and site where you are working. Refer to the PPE section in Chapter 1.

TOOLS AND EQUIPMENT

Brick trowel	Straight edge
Lump hammer	Jointing iron
Bolster chisel	Adjustable bevel
Spirit level	Steel tape measure
Block/pins and line	Compass dividers
Builder's square	Timber arch former and support wedges

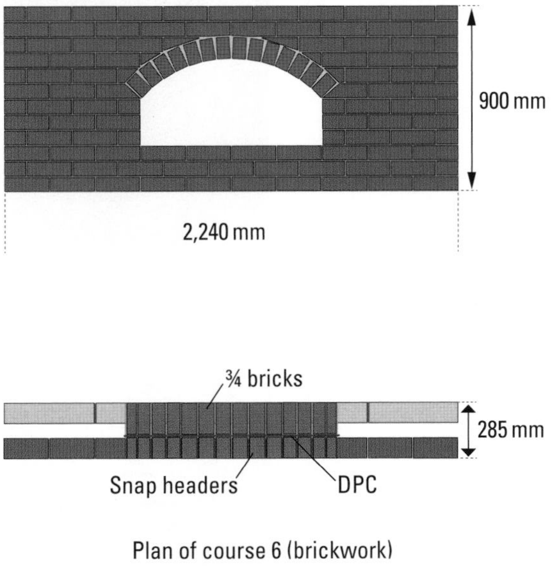

900 mm

2,240 mm

¾ bricks

285 mm

Snap headers — DPC

Plan of course 6 (brickwork)

Figure 6.53 Cross section through cavity wall with a rough ring segmental arch

STEP 1 Follow **Steps 1** to **4** of the previous practical exercise, Build a cavity wall with a rough ring semi-circular arch, on page 206.

STEP 2 Build the brickwork corners up as far as possible while allowing for the bricks that need to be cut for the skewback.

Figure 6.54 Levelling the centre

STEP 3 To form the skewback, place a brick at the end of the arch former and using an adjustable bevel, set the angle.

Transfer this to the brick that will form the skewback with a pencil, then cut and lay it.

Figure 6.55 Forming the skewback

PRACTICAL TIP

Use 10 mm packers to position the brick to be cut at the right height.

STEP 4 Follow **Steps 6** to **12** of Build a cavity wall with a rough ring semi-circular arch (pages 207–8).

Figure 6.56 Forming the skewback

Figure 6.57 Filling in bricks to make the arch

Figure 6.58 Marking curve on a brick to cut

Figure 6.59 Insert DPC and weep holes

Figure 6.60 The finished wall

Figure 6.62 The finished wall: back

Figure 6.61 The finished wall: front

TEST YOURSELF

1. In a cavity wall what creates a barrier for water?

   a. Outer wall

   b. Inner wall

   c. Window

   d. Cavity

2. What are the two ratings used for engineering bricks?

   a. A and B

   b. A and C

   c. A and D

   d. B and C

3. What is the object that is placed above the opening of a wall to support brick or blockwork above it?

   a. Jamb

   b. Sill

   c. Lintel

   d. Block

4. How many bricks are there over an air brick and liner?

   a. 1

   b. 2

   c. 3

   d. 4

5. What is the minimum width of a cavity according to Building Regulations?

   a. 50 mm

   b. 75 mm

   c. 450 mm

   d. 900 mm

6. Which of the following is used for insulating cavity walls?

   a. Mineral wool

   b. Beads or granules

   c. Insulation foam

   d. All of these

7. What sort of decorative feature can be described as being a course of cut bricks laid at 45° to the main face of the wall, which appear to be triangular in shape?

   a. Dentil

   b. Band

   c. Quoins

   d. Dog tooth

8. What are wall ties made from?

   a. Iron

   b. Wood

   c. Non-ferrous metal or stainless steel

   d. Concrete

9. To what depth should you rake out joints before pointing?

   a. 12 to 18 mm

   b. 10 to 12 mm

   c. 12 to 25 mm

   d. At least 25 mm

10. What is the upper part of a window or door frame called?

    a. Jamb

    b. Sill

    c. Rebate

    d. Head

Chapter 7

Unit CSA–L2Occ70
CONSTRUCT MASONRY CLADDING

LEARNING OUTCOMES

LO1/2: Know how to and be able to prepare for constructing masonry for framed buildings

LO3: Know how to construct masonry for use with framed buildings

LO4: Be able to construct masonry cladding to the given specification

INTRODUCTION

The aims of this chapter are to:

* help you to select materials, components, tools and equipment

* help you to apply masonry cladding to timber, steel and concrete-framed buildings.

PREPARING TO CONSTRUCT MASONRY FOR FRAMED BUILDINGS

Figure 7.1 Modern multi-storey buildings tend to have steel frames

The essential skills needed to construct masonry for either timber-framed or concrete and steel-framed structures are the same. However, bricks or blocks are not always used, and cladding has broader options.

Certainly brick or block is not ideal for tall buildings because of issues like wind, pressure, settlement and shrinkage. Therefore in some cases brick panels, which are essentially prefabricated brickwork, are used. Brick slips are an alternative; these are fixed to the building using either an adhesive or a fixing method using joints. Buildings can also be clad with panels made of concrete, stone or slate.

Traditionally, however, masonry cladding for most buildings requires that an inner and outer leaf is constructed. This will usually have a gap of between 50 and 100 mm. The two leaves are held together with ties. Effectively this is cavity wall construction. This is the normal method for buildings up to three storeys high. Taller buildings will have a cavity, but the outer skin of brickwork is tied to the main structure.

Masonry cladding can be used for both timber- and steel-framed buildings. Timber-framed buildings have become common as they are relatively quick to put up and are delivered on site as ready-made panels. These panels are then assembled. It is more common for steel frames to be used for industrial or commercial buildings and certainly for multi-storey buildings.

Hazards associated with building masonry cladding

Refer back to Chapter 1 where we examined the importance of identifying potential hazards and the ways in which risk assessments and method statements can help to avoid possible health and safety hazards. It is also important to make sure that manufacturers' instructions are followed. All of these precautions help to ensure that you follow health and safety legislation and keep yourself safe.

Drawings

When you are setting out masonry structures you will need to refer to different types of drawings.

Drawings, their purpose and how to interpret them are covered in detail in Chapter 2.

Suitable construction and specification

On-site specifications are available for most jobs and are used along with plans and drawings. Ideally the specification will give you much of the information you need.

Interpreting drawings

Details about drawings and interpreting them can be found in detail in Chapter 2.

Work method statements

To remind yourself about method statements, refer to Chapter 1.

Safety standards and procedures

To remind yourself about safety standards and procedures, refer to Chapter 1.

You should always carry out a risk assessment before starting work. Ultimately your employer is responsible for the provision of risk assessments. Assessing the specific risks first will direct you as to the best working practice for the particular job.

Personal protective equipment

Further information about personal protective equipment in general can be found in Chapter 1.

Tools, equipment and fixings

The detail and advice on tools, equipment and fixings in Chapter 5 also cover building masonry cladding.

The traditional way of building cavity walls means creating two separate leaves (inner and outer) of either brick or blockwork, with a gap between the pair of them. The two leaves are held or tied together using steel ties.

This works well for buildings of up to three storeys but for taller buildings, it is necessary to come up with a solution to provide a brickwork outer skin to cover the timber, metal or concrete frame. For example, precast concrete sections can be slotted together. These need to be lifted into position. The main point is that the outer skin is not load bearing. This is known as masonry cladding. It is the inner skin that is the load-bearing one. It can cover the following:

* Timber-framed buildings – the inner leaf consists of panels made from structural timber and these support both the roof and the floors.

* Steel-framed buildings – essentially the same system, but used mainly for commercial buildings. The cladding is tied to the main structure using metal fixings.

* Concrete buildings – these are multi-storey structures and any cladding is tied to the main load-bearing structure using metal ties.

Bricks and blocks are the most common type of cladding with the construction methods being almost exactly the same as for cavity walling. The bricks and blocks are tied to the framework using ties. The problem is that they are not capable of supporting their own weight above more than about three storeys. In addition to this, they can shrink, and settlement or wind pressure can add strain and stress.

As the building industry starts to recognise the advantages of masonry cladding, new innovations are being developed. Among the main cladding alternatives are precast brick panels or slip bricks; these are far thinner than usual bricks or blocks, and do not use traditional jointing or mortar courses. They look like bricks or blocks but are actually ready-made panels constructed in factories.

The second alternative is to use brick slips. Basically these are panels that can be attached using adhesive mortar and then finished with pointing mortar. They can be fixed to a panel, which is then fixed to the main structure. The fixing is usually by mechanical fixing methods, or by adhesives. A cement mortar is pumped into the joints to finish. They are extremely fast to install and do not need highly skilled bricklayers.

Figure 7.2 A precast brick panel

DID YOU KNOW?

Many manufacturers and suppliers offer precast panels. They consist of composite precast concrete and masonry, which can look like brickwork or natural stone.

There are also cladding systems, such as Brikloc. These are thin brick slips that are attached to galvanised steel panels that have been fastened to the structure using an adhesive or mortar. They are lightweight and extremely fast to install.

Protecting materials, completed work and surrounding area from damage

In Chapter 8 you can find information dealing with the protection of any materials you are using, completed work and the surrounding area or environment.

One of the major tasks is to ensure that adverse weather conditions do not damage materials or completed work. Equally it is important to make sure that you keep your work areas clean and deal with waste, which could otherwise affect the local environment.

Preparing and cutting components by hand or mechanical means

You may need to use a hammer and bolster chisel or a hand saw, but a disc cutter or table saw gives a cleaner finish.

Figure 7.3 Stone cladding being applied to an existing building

CASE STUDY

LAING O'ROURKE

Overcoming tough weather

Marcus Chadwick is a bricklayer at Laing O'Rourke.

'You're not supposed to wall brick when it's under 2 or 3°C, because the water freezes in the mortar, and you end up having to take it all down again. But we have ways and means of carrying on now and we've learnt from our mistakes in years gone by. When it comes to laying bricks in wet weather, we have it all tented in now. Back when we were building some schools recently, we started walling brick in December. We had all the scaffolding set up for us and then we put a big tent up with gas heaters so we could progress with the brickwork and keep the job moving without risking the quality of the bricklaying.'

REED TIP

It is useful to have a driving licence, especially if you want to be taken on permanently. You may need to drive vans and trucks.

Figure 7.4 External cladding

CONSTRUCTING MASONRY FOR USE WITH FRAMED BUILDINGS

Framed buildings are somewhat different from traditional brick and blockwork structures. The structural framework can be made from wood, concrete or steel. The load of the building is transmitted to the foundations through either a base plate or a common sole plate.

As with any new or updated structure, framed buildings have to comply with the Building Regulations; in the case of timber, particularly in relation to fire.

A big advantage of framed buildings is that much of the manufacture of the frame takes place in factories. Then the elements of the frames can be assembled on site. The finished building has better insulation, as the use of low thermal capacity linings means that the walls absorb less heat. This in turn means that the temperature inside the building rises more quickly and so uses less energy in its heating system.

The choice of external cladding is huge – from brick or blockwork to panels secured across a cavity into the frame using flexible metal ties. The wall ties are extremely important, as they allow differential movement. Over time, and depending on weather and temperature conditions, the frame and the cladding can move. As they are made from different materials from the wall, and are of different weights and properties, they may have differential movement.

Many of the points covered in earlier chapters apply to masonry cladding. Table 7.1 covers some additional points.

Activity/Product	Advice
Positioning and locating components	It is good practice to have all of the materials that you need for immediate use close at hand. You should certainly try to have sufficient components to last you for a set period of work, such as a half or full day.
Transferring vertical datum points from temporary benchmarks	Setting out a building needs to be carried out as accurately as possible. See Chapter 4 for more information.
Positioning ranging lines onto profiles and marking walling positions	Ranging lines are important when setting out. The same procedures are used as when building solid walling.
Damp-proof barriers	It is important to remember that in the case of timber-framed buildings, any timber that is below the damp-course level has to be treated for fungal or insect attack. Normally the timber that is used for frames has already been treated with preservatives.
Bonding arrangements	An additional complication is the arrangement of wall ties connecting the external cladding to the frame. This needs to be incorporated and keyed into the bonding.

Activity/Product	Advice
Forming openings	The section on forming openings in Chapter 6 deals with the forming of openings in some detail.
	All of the key considerations apply for masonry clad buildings as they do for thin joint masonry structures.
Cavity trays	A cavity tray is a shaped damp-proof course positioned at the lintel. See Chapter 6 for more information.
Insulation	The masonry clad building is still in effect a cavity wall, as there is a separation between the outer wall and the structure of the frame. The advice in Chapter 6 relating to cavity walling structures is also relevant to framed structures.
Vertical movement joints	Exactly the same precautions and conditions apply here as they do for solid walling.
Straight lengths and returns	If traditional brick or block masonry is being used as the outer cladding then the steps and practical exercise detailed in Chapter 7 should apply.

Table 7.1 Advice for installing masonry cladding

PRACTICAL TASK

FIX CLADDING TO A TIMBER FRAME WITH A RUSTICATED CORNER

OBJECTIVE

To build a section of half brick cladding onto a timber frame.

Ensure you select PPE appropriate to the job and site where you are working. Refer to the PPE section in Chapter 1.

TOOLS AND EQUIPMENT

Walling trowel	Builder's square
Lump hammer	Straight edge
Bolster chisel	Jointing iron
Spirit level	Steel tape measure
Pins and line	Boat level
Drill	Temporary timber framework

STEP 1 Set up the temporary timber frame. Make sure it is secure and will not tip (see Fig 7.5).

STEP 2 Follow **Steps 1** to **6** from the practical exercise, Build a cavity wall with return, on pages 242–3, 'or rusticated corner', on page 189. Note the different length of this model.

Figure 7.5 Setting up the temporary frame

STEP 3 Build a small external corner fixing the wall ties as per the Building Regulations to the timber frame as necessary.

Figure 7.6 Fixing wall ties to the timber frame

STEP 4 Build another corner at the stopped end. Again, fix wall ties as per the Building Regulations.

Figure 7.7 Building a corner at the stopped end (1)

Figure 7.8 Building a corner at the stopped end (2)

STEP 5 Run in the brickwork up to the window level. Again, fix wall ties as per the Building Regulations.

Figure 7.9 Running in the brickwork

STEP 6 Mark out the position of the window and continue as before until you have reached the top of the window.

Figure 7.10 Building around the window

Figure 7.11 Checking plumb around window opening

Figure 7.12 Rusticated corner in progress

STEP 7 Place the lintel over the opening and screw it in place onto the timber frame

Figure 7.13 Fixing the lintel over the opening

Figure 7.14 Building over the lintel

STEP 8 Continue to build up the remaining three courses.

Figure 7.15 Building up the remaining three courses

STEP 9 Check the width of the opening with a tape measure to make sure that the soldier course will fit. You may need to make the joints larger or smaller to make sure the soldier fits with evenly spaced joints. Ideally the opening should be 910 mm as then the bricks will fit correctly with a 10 mm joint. If not, the joints should be adjusted accordingly.

STEP 10 Some bricklayers will use two lines for walling in the soldier course, one at the top and one about two-thirds of the way up from the bottom. Keep checking each brick with a boat level for plumb.

Add weep holes after the first soldier from each end and place one in the middle of the opening.

PRACTICAL TIP

Do not build the two sides of the window up separately or the sides of the window could end up out of line with each other and on longer walls it will end up looking twisted.

PRACTICAL TIP

Always select the bricks for the soldier course. They should be the same size and have good straight edges on the face of the brick. This will make it easier to lay the soldier course.

Figure 7.16 A cavity fire barrier

Provision of fire barriers

The performance of fire barriers largely depends on the stability, or integrity, of the cladding system. If there is a distortion in the cladding then the gap holding the fire barrier may be wider and it will lose its integrity. The way around this is to make sure you use steel brackets. These prevent the separation between the frame and the cladding.

The thickness of the cavity fire barrier is measured in terms of the distance between one compartment and the next:

* 75mm to 600mm cavities will have a fire resistance of 60 minutes

* 100mm with lap joints will have a fire resistance of 120 minutes

* 100mm extra high density barrier with lap joints will have a fire resistance of 240 minutes.

The fire barrier is usually made from rockwool. The purpose of the material is not only fire protection, but also to stop any smoke from spreading. The material also has thermal and acoustic (noise reduction) properties.

The fire barrier stops, or at least inhibits, flames, heat and smoke from moving through concealed spaces in the building.

TEST YOURSELF

1. What is the usual width of the gap between the inner and outer leaves of a cavity wall?

   a. Between 25 and 50mm

   b. Between 50 and 100mm

   c. Between 100 and 120mm

   d. There is no usual width

2. Which type of building is most likely to have its basic construction made from steel frames?

   a. A domestic dwelling

   b. An extension

   c. A garage

   d. A commercial building

3. In multi-storey concrete buildings, what is used to secure the cladding to the main load-bearing structure?

   a. Metal ties

   b. Concrete slots

   c. Pre-cast clips

   d. Bolts

4. What is the purpose of low thermal capacity linings?

   a. The linings heat up quicker

   b. The linings are constructed using low temperature processes

   c. The walls absorb less heat

   d. The linings are damaged if they become overheated

5. Wall ties allow differential movement. What causes differential movement?

   a. The cladding can be made from a different material to the wall

   b. Cladding can be of different weights

   c. The cladding can have different properties

   d. All of these could cause differential movement

Chapter 8

Unit CSA–L2Occ71
CONSTRUCT THIN JOINT MASONRY

LEARNING OUTCOMES

LO1/2: Know how to and be able to prepare for constructing thin joint masonry to the given specification

LO3: Know how to set out for constructing thin joint masonry to the given specification

LO4/5: Know how to and be able to construct thin joint masonry to the given specification

INTRODUCTION

The aims of this chapter are to:

* help you select materials, components, tools and equipment

* show you how to construct buildings using thin joint masonry.

PREPARING TO BUILD THIN JOINT MASONRY TO A GIVEN SPECIFICATION

Thin joint masonry, or blockwork, has allowed the construction industry to build higher quality buildings much faster. The mortar used is a pre-mixed, special thin joint masonry adhesive product, which only needs water to be added. The key benefits of using thin joint mortar and **aerated (aircrete) blocks** include:

* faster build speed

* increased productivity, as the mortar is usually stable after 60 minutes

* more accurate construction, using 2–3 mm joints

* improved thermal performance, as blockwork reduces the amount of mortar being used by around 70 per cent

* good air tightness

* good sound insulation

* high fire resistance

* reduced site waste, as the blocks can be sawn or cut on site

* improved construction quality, as the blocks provide an accurate internal face that only needs a thin coat of plaster

* flexibility – the technique can be used for foundations, partition walling, external solid walling, cavity walling and separating or party walls.

Health and safety hazards

In Chapter 1 we examined the importance of identifying potential hazards and the ways in which risk assessments and method statements can help to avoid possible health and safety hazards. It is also important to make sure that manufacturers' instructions are followed. All of these precautions help to ensure that you follow health and safety legislation and keep yourself safe.

Figure 8.1 Bricklayer working on a thin joint wall.

Drawings

When you are setting out masonry structures you will need to refer to different types of drawings, but you will also have to understand:

* abbreviations
* hatchings
* scales used.

Each type of drawing has a specific purpose and together they should provide you with the full picture of exactly what is required.

The purpose of drawings is to assist construction. They are organised in a logical sequence, which should follow the flow of the actual construction work. All details and sections are usually grouped together on the same sheet. Scales should be clear and show the details of the structure.

Drawings and their purpose are covered in detail in Chapter 2.

Suitable construction and specification

On-site specifications are available for most jobs and are used together with plans and drawings. Ideally the specification will give you much of the information you need and determine the way in which you construct the blockwork:

* It will state the type and size of all materials, including the quality of the materials.
* It will detail the types of fixing and finishing required.
* It will tell you what services are available.

On smaller jobs, specifications may not be available but they are common on most construction sites.

Interpreting drawings

All drawings must follow the requirements of BS 1192:2007. This means that the drawings will have a common format and symbols. Building drawings use what is known as first angle orthographic projection. Drawings in this form will have a special symbol.

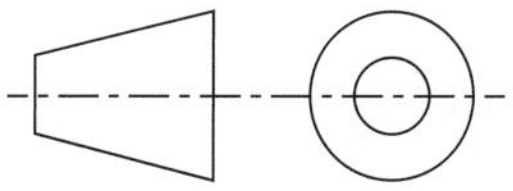

Block and site plans and general location

Block plans are drawn at 1:1250, or 0.8 mm to 1 m. Site plans can be drawn either at 1:500 or 1:200, which are either 2 mm or 5 mm to 1 m. General location plans can be variously drawn at 1:200, 1:100 or 1:150, which are 5, 10 and 20 mm to 1 m.

Figure 8.2 First angle projection symbol

Details and assembly

Details can be drawn at 1:10, 1:5 or even 1:1. These are 100 mm or 200 mm to the metre, and in the case of 1:1 they are full size. Assembly drawings are at 1:20, 1:10 or 1:5, which means they are 50, 100 or 200 mm to 1 m.

Sectional

Section lines provide you with vertical dimensions and construction details. They also cover:

* damp-proof courses and membranes

* foundations, floors or walls

* roofs.

These also give the height of ground levels. See Fig 2.4 for a section drawing of an earth retaining wall.

First angle orthographic projection

First angle projection is a view that represents the side of the object remote from it in the adjacent view.

Isometric projection

Isometric projection is a way of representing three-dimensional objects in two dimensions.

Work method statements

According to the Health and Safety Executive a work method statement is an ideal way to record any specific hazards that might be involved in carrying out work. The method statements are clear and often illustrated with sketches. These statements are for the benefit of those who are carrying out the work, and are not complicated. They will also state the type of equipment needed.

Safety standards and procedures

We have already seen in Chapter 1 that a construction site can be a very dangerous place. In all types of construction work there are potential hazards. It is very important to identify the risks and then work in a safe way in order to minimise them.

Each different job will need a different combination of tools, materials and equipment. It is essential to take particular care when cutting and laying blocks. Site safety rules should be followed at all times.

You may also be working with substances, such as cement, which are considered to be hazardous to health. These hazardous materials will always have data sheets to tell you exactly how to use them and any potential hazards.

You should always check the risk assessment provided by your employer before starting work. Assessing the specific risks first will direct you as to the best working practice for the particular job. There are several health and safety issues, which are outlined in Table 8.1.

Health and safety issue	Solution
Mortar splashing onto the skin	Wear PPE – particularly gloves and safety goggles.
Mortar dust and fumes	Wear PPE as above and a respiratory mask.
Block unloading and carrying	Always unload blocks from delivery truck using mechanical aids and use these to transport the blocks on site.
Blocks and scaffolding	Use mechanical aids to get the blocks to the correct level and do not carry them by hand up ladders.
Exposed wall ties	Never leave these exposed, as others may walk into them or catch their clothing on them.
High blockwork	Make sure that blockwork is properly braced and protected if working at height. Also, workers working at height must not have to strain to reach things, and must be protected from falling.

Table 8.1 Health and safety issues

Personal protective equipment

PPE may include:

* safety goggles or safety glasses – always necessary if there is a risk of injuring or getting things in your eyes

* gloves – these protect skin from mortar and other potentially damaging substances and against getting dermatitis, help grip and protect hands if handling sharp objects

* safety footwear – needed in case of dropping heavy items, standing on sharp objects, uneven ground and poor weather conditions

* hard hat – if it is possible that items could strike your head.

The required PPE will depend on the risk assessment, although all of the listed PPE is compulsory on some of the larger construction sites.

Hand tools and equipment

You must use tools specifically for thin joint walling rather than standard bricklaying tools. You need to use the proper application tools to ensure you apply the right amount of mortar to the surface bed of the block and the cross joint:

* A scoop or sledge per block layer – to spread thin joint mortar. This will produce a consistent 2–3 mm thickness.

* A sanding board, which can be shared by two block layers – to remove imperfections in the bed course.

Figure 8.3 Materials and equipment for mixing thin joint mortar and for the subsequent construction process

* A block rasp, which can be shared by two block layers – to trim areas of the block that are raised which cannot be removed using a sanding board.

* A block cutting square – to be used as a marking guide when cutting blocks.

* A bucket or large tub – to hold mortar.

* A whisk attachment for a drill – to mix mortar.

You also need a hand or bench saw to cut the blocks.

Other key pieces of equipment include a spirit level, damp-proof barriers and wall ties.

You need to use special wall ties to join the two leaves of a cavity wall because the aircrete blocks don't match up with the brick leaf as they would in standard walling. Helical ties can be driven into the face of the blockwork at a level which will meet with the outer leaf mortar joint. Insulation can be placed against the inner leaf before wall ties are inserted.

You position the ties as the outer leaf is being built, to prevent waste mortar building up on the ties and to reduce the risk of injury. You can also lay ties in the thin joint beds as you lay the blocks, especially when the inner and outer leaves are being built together.

Protecting materials, completed work and surrounding area from damage

Materials

Safe storage of materials, such as cement, means:

* not puncturing the bags before use

* storing them on pallets in a waterproof and ventilated area

* not piling the bags more than five high

* using the bags in the order in which they were delivered.

Blocks, like bricks, are delivered by pallet and have bands around the edges and strips to hold them in place. They are also usually covered in plastic. They should be stored no more than two pallets high and be as close to where they are needed as possible. Remember that each time you move blocks you risk damaging them.

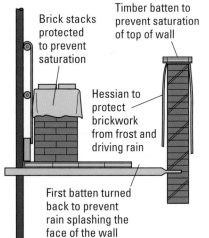

Figure 8.4 Protection of a wall

Completed work

Newly built blockwork can be vulnerable to cold temperatures and rain. The blockwork should be encouraged to dry out, which may mean putting a covering material over the face of a wall. Hessian is also used as an insulating layer.

Cold weather can weaken walls by causing the water in the mortar to freeze. A hessian layer will act as a good insulator.

The blockwork should be encouraged to dry out, and any coverings should be kept clear of the wall to allow for ventilation. Coverings should be secured using a scaffolding board or bricks to prevent it from being blown away.

Surrounding area

Waste is unavoidable but strict environmental legislation determines how you handle and dispose of waste. The waste that is produced will cause some environmental damage and this waste must be minimised by:

* reusing broken bricks and blocks as hardcore

* sweeping fine debris into heaps and sprinkling water on it to minimise the dust

* making sure that unused mortar is not remixed as this is bad practice

* not burning waste

* keeping all material bins shut or locked when not being used

* bagging up any waste, or at least covering it up.

Preparing and cutting components

It may be necessary for you to cut and prepare the blocks with a hand saw or bench saw – this method is only possible with lightweight blocks as the hand saw won't be strong enough to cut through dense blocks.

PRACTICAL TIP

Some parts of the UK are exposed to severe weather conditions. These are areas that have higher than average rainfall, more frost or are over 90 m above sea level. Special precautions need to be taken in these areas.

CASE STUDY

Bricklaying is just the start

Marcus Chadwick is a bricklayer at Laing O'Rourke.

'One of the reasons I enjoy brickwork is because we build iconic structures – some have even won awards for design. I've worked on train stations, museums, hospitals, schools, and more. Because we don't do as much brickwork as we used to, we've had to diversify and learn other skills, for instance we'll do the concreting for slabs, and then we'll go out and do the kerbing, the flagging, the block paving.

I've done quite a bit of further training – my NVQ Level 3, and in Plant Operations too. I've got tickets to drive a lot of the machinery on site, again because I've had to diversify from just bricklaying. So now I'll drive forklifts, rollers … you name it, I can drive it. I've also done first aid, advanced first aid, scaffolding, and inspecting scaffolds. If you show an interest in wanting to progress further in life, then this company will give you that little helping hand and the confidence to do it, even if it's for higher education. I'm quite happy being a supervisor, though I once had the inkling to progress further, but you've got to see what your workload is like – for NVQs 4, 5 and 6, it's quite a lot. Also, I like my site work, I like to get my hands dirty, and it wouldn't suit me to sit in an office. I'd sooner be out in the cold and fresh air.'

LAING O'ROURKE

PRACTICAL TIP

Obviously where you place the components may have an effect on others. You may block access with your components, or you may create a health and safety hazard. So always check and be sure.

SETTING OUT A BUILDING FOR THIN JOINT MASONRY

Setting out procedures should follow the same process that we covered in detail in Chapter 4. There are some specific guidelines that you should also follow, which are outlined in this section.

Positioning and locating components ready for use

Before starting the job it is advisable to stack all the components you need to work with. It is not simply a question of putting them as close to the work as possible. Each building site may have different procedures. Some will encourage placing the components close by, but on others they may have to be in a compound until they are needed.

You should make sure that:

* you follow any written instructions about the storage or placement of components

* if there are no written instructions you seek verbal instructions from the site manager

* you check with the site manager if you are unsure about where the components should be placed.

Position ranging lines onto profiles and marking walling positions

Refer to Chapter 4 for more information on ranging lines and profile boards. The figures below show how they are used in setting out.

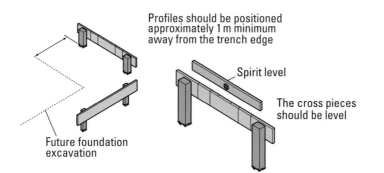

Profiles should be positioned approximately 1 m minimum away from the trench edge

Spirit level

The cross pieces should be level

Future foundation excavation

Figure 8.5 Profile boards set back from building

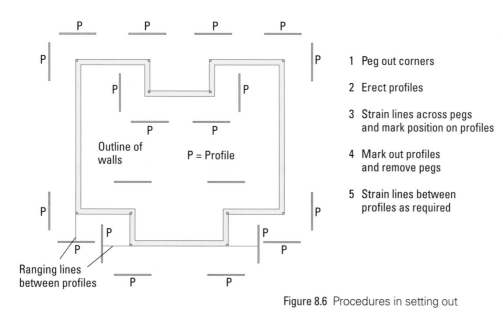

Outline of walls

P = Profile

Ranging lines between profiles

1 Peg out corners

2 Erect profiles

3 Strain lines across pegs and mark position on profiles

4 Mark out profiles and remove pegs

5 Strain lines between profiles as required

Figure 8.6 Procedures in setting out

Establishing bonding arrangements

Refer to Chapter 5 for information about bonding.

Maintaining industrial standards

As with all types of blockwork walling, thin joint system should be built according to British Standard Codes of Practice for the use of masonry covering materials, components, design and quality.

The Concrete Block Association and manufacturers are useful sources of advice. Consider the following:

* You must set out the base or first course correctly as mistakes are hard to correct later. Bed base course blocks in general mortar, then level, align and plumb and allow to set before starting thin joint construction.

* Plan for any damp-proof course to be incorporated into the base course bed joint.

* Regular bond pattern block laying should always be the goal and there should always be an overlap of at least a quarter of a block.

* You should always use cut blocks for irregular spaces rather than other materials.

* Blocks should be 3 mm apart for thin joint blockwork.

* Lintels should always be set on full blocks.

CONSTRUCTING BUILDINGS USING THIN JOINT MASONRY

The techniques you need to use for thin joint walling are similar to those for general mortar. You still need to check level and line regularly. For cavity walls, you can use a mixture of standard, large format and cut blocks to meet datum levels for floor and wall levels.

Transferring horizontal and vertical datum points

It is vital that whenever a building is going to be constructed it is set out properly. Lines and levels must be accurate. The survey covers both the **horizontal** and the **vertical** levels. See Chapter 4 for more information about transferring datum points and levels.

See Chapter 6 for information on cavity trays and insulation, and Chapter 7 for information on damp-proof barriers and forming openings.

Advantages of thin joint walling

Using thin joint systems speeds up construction in several ways:

- Mortar can be applied very quickly, even before the blocks have been unpacked.

- In good weather, the mortar mix stays workable for up to 30 minutes before blocks are laid. Once they are laid, mortar sets within 10 minutes.

- You can build the inner leaf of cavity walls first, which not only makes it easier to install insulation but also means that internal work like first fix carpentry and services can go ahead before the outer leaf is built.

Reinforcement

Thin joint reinforcement is usually achieved by using either galvanised steel or stainless steel wire. The reinforcement is bedded into the mortar. There are guidelines how this should be achieved:

- Any overlap should be a minimum of 225mm.

- Around openings, reinforcement should extend by at least 600mm on either side of the opening.

- This type of reinforcement can also be used instead of movement joints for blockwork that is over 6m.

- The reinforcement is built into the horizontal bed of the thin joint wall. This helps to protect the blockwork from tensile forces.

Mixing jointing compounds

Usually a special pre-mixed jointing compound is used and you need to mix this with water.

The mortar will begin to set about half an hour after the water has been added. So this means that only relatively small quantities of mortar should be mixed in one go.

See page 159 for information on vertical movement joints.

Jointing and pointing

Jointing will depend on the desired look of the wall. The timing of jointing will depend on a number of factors:

- Moisture – if the bricks are wet then you will have to wait to joint until they have dried out.

- Temperature – on hot days joints will dry out quickly, as the bricks will absorb heat. On colder and wetter days joints may take more time to dry out, as will the bricks.

- Time of day and progress of work – brick courses that you have already laid earlier in the day will be drier than those that you have worked on more recently.

PRACTICAL TIP

Try touching the joint to see how dry or wet it is. If the joint is too wet then the mortar will drag and the effect will be quite messy.

PRACTICAL TIP

If you are using an angle grinder make sure you do not allow the grinder to pass over the face of the brick, as this will damage it. Also angle grinders create a lot of dust, which can become a problem.

* Type of brick – the softer the brick, the more moisture it is likely to soak up, which in turn means that the joints will take longer to dry out.

Before pointing can take place joints have to be raked out, usually to a depth of between 12 and 18 mm. There are two ways of doing this:

* For soft joints use a raking out tool.
* For drier joints, and for bricklayers with sufficient experience, an angle grinder can be used.

In many cases a combination of the two can be used, but you can only begin to point once the raking has been completed.

Pointing is the process of filling the mortar joints. There are several different ways in which the desired effects can be achieved. You should remember that pointing needs to be carried out slowly and carefully as this will be the finished look on the wall.

The following advice should always be followed when pointing:

* Always start your pointing at the top of the wall. This is to ensure that any loose mortar does not fall onto freshly pointed areas.
* Ensure that the raking out has been done properly and that there is no loose mortar or dust in the joint.
* Remember to dampen the bricks around the area that you are pointing. You need to bear in mind that different bricks absorb water in different ways. Ensuring that the bricks are dampened will mean that your mortar will stick more easily.
* Once you have completed a section of pointing and it is dry, use a brush to clean off any dust.

Joint finishes

There are at least six different types of joint finish that you can achieve when pointing. These can be seen in Table 8.2.

Joint finish	Description
Weather struck	The joint slopes downwards to encourage rainwater away from the joint and down the face of the brick.
Ironed or tooled	One of the quickest jointing finishes – a shallow semi-circular indent is made in the joint.
Recessed	During pointing the mortar is compressed into the joint to a depth of around 4 mm.
Flush	The mortar completely fills the joint up to the face of the brick.
Weather struck with cut pointing	The joint angles downwards and protrudes out beyond the face of the brickwork.
Reverse struck	This is the complete opposite of the weather struck joint. The joint angles inwards from the top. It is not normally used outdoors because water will stand on the lip of the brick, which weathers it.

Table 8.2 Different types of joint finish for pointing

Installing components

The term component refers to all of the additional features that are necessary in building construction. These include:

* the blocks themselves
* frames
* insulation
* damp-proof barriers
* lintels
* fixings and ties.

The drawings and specification will detail precisely what is required, which components are necessary and their location and size.

Returns and junctions

It is important to maintain the half bond at junctions and returns. This will mean cutting blocks.

Fig 8.8 shows the 100 mm cut block being used. An alternative method requires additional cutting and in effect it means having additional cut blocks on every other course at corners.

Junctions require at least a quarter of a block overlap. Cut blocks of 100 mm can be used alternately in courses.

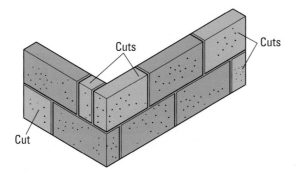

Figure 8.7 Block return corner

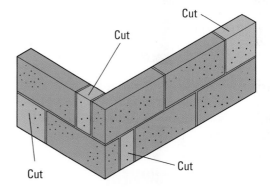

Figure 8.8 Alternative block corner

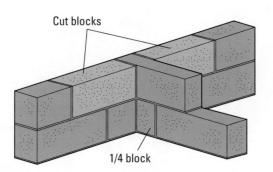

Figure 8.9 Junction position in centre of block in main wall

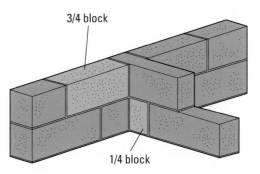

Figure 8.10 Junction position off centre reduces the amount of cutting

1. BUILD A 140 MM BLOCK WALL WITH RETURN AND OPENING

OBJECTIVE

To construct a brick and 140 mm block cavity wall with a return and an opening, using the techniques required to build a model in thin joint masonry.

With this system, the joints are only 3 mm, which means that accuracy when setting out is crucial. The time spent at the start pays for itself with quicker build times for a given area of masonry. Because the complete inner shell of a house can be built in thin joint masonry, other trades can continue working inside the structure while the outer leaf is being formed.

The measurements required are:

Length of wall 2,650 mm

Return length 660 mm

Width of window opening 885 mm

Thin joint blocks 440 mm × 215 mm × 140 mm

Lintel steel box section 140 mm × 140 mm × 1,200 mm

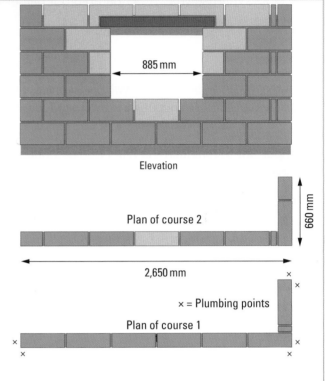

Figure 8.11 A 140 mm block wall with a return and opening

Ensure you select PPE appropriate to the job and site where you are working. Refer to the PPE section in Chapter 1.

TOOLS AND EQUIPMENT

Brick trowel	Thin joint spreaders
Spirit level	Sanding rasp
Block/pins and line	Cutting square
Straight edge	Block saw
Steel tape measure	Builder's square

STEP 1 Mix the mortar in accordance with the manufacturer's instructions until the desired consistency is reached.

STEP 2 Dry bond the blocks to the required length of 2,650 mm.

Figure 8.12 Setting out wall

Figure 8.13 Transferring levels (1)

Figure 8.14 Checking first course for plumb

STEP 4 Run the blocks in to complete the first section of the wall using the thin joint mortar on the perp joints and conventional mortar on the bed joints.

Figure 8.16 Running blocks into first level

Figure 8.17 Checking the wall is level

PRACTICAL TIP

Always use the cutting guide when marking out blocks to be cut. This ensures that the cuts will fit correctly with the thin joint mortar, instead of leaving large voids.

STEP 3 Bed the end blocks on normal mortar beds at either end of the wall. Then check for plumb, level and alignment as well as for the length of the wall.

Figure 8.15 Applying jointing compound to block and spreading it

STEP 5 Lay the blocks on the return, ensuring that the wall is the accurate length (660 mm) as well as checking for level, plumb and that relevant angles are 90°.

Figure 8.18 Marking and cutting block to size

Figure 8.19 Laying the blocks on the return and ensuring plumb and level (1)

Figure 8.20 Laying the blocks on the return and ensuring plumb and level (2)

STEP 6 Build the corners up to the required height, making sure that the spacing for the opening on the third course is correct. An alternative method is to construct the second course and mark the window opening before the corners are built up to height.

Figure 8.21 Running out second course and checking alignment

Figure 8.22 Smoothing out high spots

Figure 8.23 Marking in the window with mortar

Figure 8.24 Marking out the window with pencil

Figure 8.25 Building up corners and checking plumb

STEP 7 Run the intermediate blocks in while checking that the sides of the window are plumb with the level or by using a dummy window frame.

Figure 8.26 Checking the window is plumb

PRACTICAL TIP

Use the string line for alignment only, as the system means that the height is maintained by the thickness of the bed joints.

STEP 8 Bed the lintel on at the required height with conventional mortar, and check dimensions for accuracy.

STEP 9 Cut blocks around the lintel, making sure that any blocks bedded onto the lintel are placed onto conventional mortar.

Figure 8.27 Installing a lintel

Figure 8.28 The finished result

PRACTICAL TASK

2. BUILD A CAVITY WALL

OBJECTIVE

To construct a brick and 140 mm block cavity wall.

This model consists of a section of wall built in thin joint masonry. One advantage of thin joint walling is that the inner skin can be constructed independently of the outer skin. This means that a shell of a building can be completed with the bricklayers building the outer skin while other trades continue internally.

It becomes an easier process because the wall ties are fitted after the wall has been constructed. Therefore, the blockwork must be built first. While the first brick course can be laid so that the position of the inner skin can be reached, the following instructions will concentrate on the blockwork being built independently.

The measurements required are:

Length of wall 2,690 mm

Width of cavity 100 mm

Thin joint blocks 440 mm × 215 mm × 140 mm

TOOLS AND EQUIPMENT

Brick trowel

Spirit level

Block/pins and line

Straight edge

Steel tape measure

Thin joint spreaders

Sanding rasp

Cutting square

Block saw

Thin joint training adhesive mixed to a suitable consistency

Thin joint ties that are hammered into the blockwork and bedded into the outer brick leaf as work proceeds

Ensure you select PPE appropriate to the job and site where you are working. Refer to the PPE section in Chapter 1.

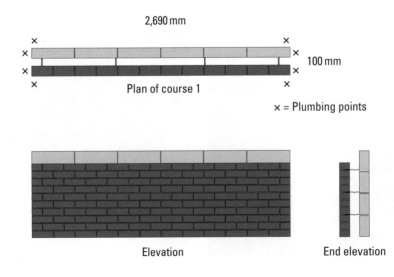

Figure 8.29 Plan of brick and block cavity wall

STEP 1 Mark the position of the front face of the wall so that the setting out procedure can be established from a given point.

STEP 2 From these marks, draw the position of the face of the inner skin so that the thin joint masonry can be built first, as it would be on site.

STEP 3 Lay the first block on the internal corner on traditional mortar and transfer this to a block placed 2,690 mm away, at the far end. Check for level and alignment and re-check the length before running in, using the thin joint mortar for the perps (vertical gaps).

PRACTICAL TIP

Ensure that all the blocks are plumb before any work continues, so that the corner's accuracy is maintained. A hammer is ideal for this job.

STEP 4 Build up the corners to full height, using the level to check for plumb only. The accuracy of the joint means that levelling should be unnecessary.

STEP 5 Run in the remaining blockwork to the line, which should be positioned to check the alignment only.

PRACTICAL TIP

If any raised edges are present, the sanding block can be used to flatten the bedding surface so that accuracy is maintained.

STEP 6 Position the quoin (corner) brick on the corner ensuring that the correct cavity width is established. Transfer the level to the other end of the wall 2,690 mm away and re-check all dimensions.

STEP 7 Build the corners up in small sections and run in to maximise efficiency until full height is reached. Use the correct tool to screw in the ties as the corners are being built.

STEP 8 Finish the brickwork with a brushed half-round joint.

PRACTICAL TASK

3. BUILD A CAVITY WALL WITH A RETURN

OBJECTIVE

To build a section of cavity wall built in thin joint masonry with a return.

The measurements required are:

Length of wall 2,690 mm

Width of cavity 100 mm

Thin joint blocks 440 mm × 215 mm × 140 mm

Ensure you select PPE appropriate to the job and site where you are working. Refer to the PPE section in Chapter 1.

TOOLS AND EQUIPMENT

Brick trowel	Thin joint spreaders
Spirit level	Sanding rasp
Block/pins and line	Cutting square
Straight edge	Block saw
Jointing iron	Builder's square
Steel tape measure	

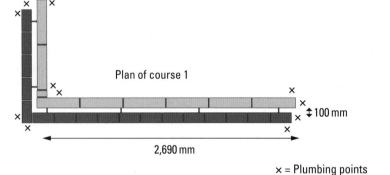

Plan of course 1

2,690 mm

✚ 100 mm

× = Plumbing points

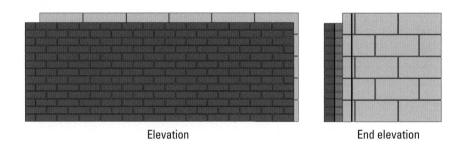

Elevation

End elevation

Figure 8.30 Plan of block wall with brick cavity and a return

STEP 1 Using either the drill, or the mega mixer with a whisk attachment, mix the adhesive in accordance with the manufacturer's instructions until the desired consistency is reached.

STEP 2 Set out and dry bond the blocks to the required length of 2,650mm.

STEP 3 Bed the end block on conventional mortar bed. Then check for gauge, plumb, level, and alignment.

STEP 4 Now lay the other end block with conventional mortar and, using a straight edge and level, make sure that it is level with the first.

STEP 5 Run the blocks in to complete the first course of the wall using conventional mortar for the bed joint and thin joint mortar for the cross joint.

PRACTICAL TIP

Ensure that all the blocks are plumb before any work continues, so that the corner's accuracy is maintained.

STEP 6 Lay the blocks on the return, ensuring that the wall is the correct length (550mm), as well as checking for level, plumb and that the angle is 90°.

STEP 7 Build up the corners, using the thin joint adhesive.

PRACTICAL TIP

If any raised edges are present, the sanding block can be used to flatten the bedding surface so that accuracy is maintained.

Always use the cutting guide when marking out blocks to be cut. This ensures that the cuts will fit correctly with the thin joint mortar, instead of leaving large voids.

STEP 8 Run in the remaining blockwork to the line, using the thin joint adhesive. Don't forget to lay the blocks for the junction wall as well.

STEP 9 Bed the corner brick to gauge at the corner ensuring that the correct cavity width is established.

Bed another brick at the other end of the wall making sure that it is 2,690mm away from the corner brick. Using a straight edge and level, level the brick with the corner brick.

STEP 10 Run out the return with a spirit level and make sure that it is level to the main wall as well as 90° to the main wall.

STEP 11 Build the corners up in small sections and run the bricks in between the corners. Knock in the twist ties as necessary.

Finish the brickwork with a half-round joint as the work proceeds.

Figure 8.31 Fixing ties

TEST YOURSELF

1. How wide are the joints in thin joint systems?

   a. 1–2 mm

   b. 2–3 mm

   c. 5–6 mm

   d. 10–12 mm

2. Which of these is NOT a benefit of using aerated blocks?

   a. Improved thermal performance

   b. Can be used to match conventional blocks

   c. High fire resistance

   d. Less site waste

3. Which is the best organisation to contact for information and advice about using thin joint systems?

   a. HSE

   b. CITB

   c. The Concrete Block Association

   d. BREEAM

4. How long does thin joint mortar take to set once blocks have been laid?

   a. 30 minutes

   b. 60 minutes

   c. 10 minutes

   d. 90 minutes

5. What is joint reinforcement usually made of?

   a. Aircrete blocks

   b. Steel

   c. A pre-mixed jointing compound

   d. uPVC

6. Why should you use the cutting guide when marking out blocks to be cut?

   a. So that the cuts will fit with the thin joint mortar

   b. To protect your hands when you a sawing the blocks

   c. Because aerated blocks are much harder to cut than standard blocks

   d. To make sure blocks match with the outer leaf of the cavity wall

7. What is a disadvantage of the thin joint system?

   a. The blocks are very heavy

   b. Any inaccuracies in the first course must be corrected straight away

   c. It is a slower form of construction than standard blockwork

   d. Its joints are only 3 mm thick

8. When might you use conventional mortar when constructing thin joint walling?

   a. To strengthen the joints in taller buildings

   b. To replace pre-mixed thin joint mortar if you run out

   c. To increase the wall's weatherproofing qualities

   d. To make the first foundation level enough to build on

9. What is an advantage of laying the inner leaf of a cavity wall first?

   a. It strengthens the walls

   b. The first fix can happen while the outer leaf is being built

   c. It speeds up the construction of the outer leaf

   d. You don't have to use wall ties

10. How do you correct raised edges on a block?

    a. Cut them off with a hand saw

    b. Sand them down

    c. Cover them with thin joint mortar

    d. Cover them with conventional mortar

INDEX

3:4:5 method, right-angle corners 110–11

A
abbreviations 50
accident books 2, 9, 17
accident procedures 8–13
additives
    foundations 72–3
    mortars 135–6
adhesives, solid walling 136
admixtures, mortars 135–6
aerated blocks 224
agendas 65
aggregates, foundations 72
alternative building methods, sustainability 87–8
alternative energy sources, sustainability 91–3
arches
    cavity walling, rough ring segmental arch 205–11
    cavity walling, semi-circular arch 205–8
    solid walling 173–5
    voussoirs 173, 174
architecture, sustainability 85–6
arris, brick term 139
asbestos 24, 101
    Control of Asbestos at Work Regulations 3
assembly drawings 42
assembly points 34

B
batt, brick term 139
batten, exterior roof feature 80
beam and block floors 74
bed joint, brick term 140
biodegradable materials 88
biodiversity 84
biomass stoves 91
block plans 41
block wall with return and opening, thin joint masonry 236–9
blockwork see thin joint masonry
bonding
    dry bonding 144
    solid walling 138–42
brick closer, brick term 139
brick slips, masonry cladding 216–17
brick terminology 138–42
bricks and blocks, cavity walling 179
Brikloc cladding system 217
British Standards 97, 121
building design, sustainability 85–6, 89, 93
building lines 107–9
Building Regulations 89, 93, 98

C
CAD (computer-aided design software) 129
carbon, energy sources 90
carbon footprint 85
cavity frames, cavity walling 180
cavity trays, cavity walling 180
cavity walling 77, 178–212
    arches 205–11
    bricks and blocks 179
    cavity frames 180
    cavity trays 180
    cavity wall with a return, thin joint masonry 242–3
    condensation 182–3
    cutting components 181

    damp-proof barriers 195–6
    damp-proof courses (DPCs) 180, 195
    decorative features 197–8
    external return and junction 187–9
    fixings 204–5
    foundations 182–3
    gable ends 192–4
    insulation 179, 182, 196–7
    lintels 180, 202–3
    openings 201–3
    party wall junctions 184–6
    piers 204–5
    return and rusticated corner 189–92
    rough ring segmental arch 205–11
    semi-circular arch 205–8
    setting out 201
    specifications 181–4
    straight lengths 183
    thin joint masonry 240–3
    vertical movement joints 198–200
    wall ties 180, 198–200
    working space 181
CCHP (combined cooling heat and power) units 92
cement
    foundations 72
    solid walling 132–3
cladding see masonry cladding
clearance activities, preparing construction sites 100, 102–3
colour coded cables, electricity 31
combined cooling heat and power (CCHP) units 92
combustible materials 18, 34–6
communication 58–65, 113
    see also information
    diversity 61–2
    documentation 62–4
    equality 61–2
    key personnel 59–60
    methods 62–5
    poor 61
    teamwork 61
    types 58–9
competent (individuals/organisations) 4
computer-aided design software (CAD) 129
concrete
    floors 75
    formwork 75
    foundations 72
Concrete Block Association 226
concrete buildings, masonry cladding 216–22
concrete coping
    English bond walls 160–4
    solid walling 160–4
condensation
    cavity walling 182–3
    foundations 182–3
contamination 18, 19
Control of Asbestos at Work Regulations 3
Control of Substances Hazardous to Health Regulations (COSHH) 2, 3, 20
coordinating sizes, bricks 139
coping, solid walling 158
corner profiles, right-angle shaped masonry structures 118–19
COSHH see Control of Substances Hazardous to Health Regulations
course, brick term 140

covering letters 65
cross joint, brick term 140
cross junction, brick term 140
cutting components, cavity walling 181

D

damp-proof barriers
    cavity walling 195–6
    solid walling 155–6
    tanking 156
    thin joint masonry 232
damp-proof courses (DPCs) 78
    cavity walling 180, 195
damp-proof membranes (DPMs) 78
dangerous occurrences 9
datum heights 120
decorative corners, solid walling 157
decorative features, cavity walling 197–8
dermatitis 20, 21, 32
detail drawings 44
dimensional accuracy, measurements 98, 113, 121
discrepancies/inaccuracies
    setting out 98
    solid walling 130
diseases 9
documentation 62–4
    see also communication; information
doors, solid walling 173–5
double pitched roofs 79
DPCs *see* damp-proof courses
DPMs *see* damp-proof membranes
drawings 41–4
    equipment 50
    interpreting 225–6
    scales 51
    setting out 97
    thin joint masonry 225–6
dry bonding, solid walling 144

E

ear defenders 20, 21, 32
ECOSHH Regulations, noise 20
electricity 28–31
    colour coded cables 31
    dangers 29–30
    Portable Appliance Testing (PAT) 28–9
    precautions 28–9
    voltages 30, 31
emergency procedures 8–13, 34–6
energy efficiency, sustainability 89–93
energy ratings 93
English bond walls 143–5, 147–50, 157–9, 160–4
    attached piers 164–5
    concrete coping 160–4
    squint corners 157–9
    two-brick hollow isolated pier 167–9
    two-brick solid isolated (detached) pier 166–7
English garden wall 151–4
environment 85
    see also sustainability
estimates, job 58, 63
estimating quantities of resources 51–8
external return and junction, cavity walling 187–9
eye protection 32

F

fascia, exterior roof feature 80
fire barriers, masonry cladding 222
fire extinguishers 35–6
fire procedures 34–6
firings, exterior roof feature 80

first aid 12–13
first angle projection (orthographic) 44
fixings
    cavity walling 204–5
    masonry cladding 215–17
flat roofs 78, 79
Flemish bond return corner, solid walling 146–7
Flemish bond T-junction, solid walling 150–1
Flemish garden wall bond 154–7
    attached piers 165–6
    attached piers and movement joint 169–70
floating floors 74
floors 73–5
    alternative building methods 88
    concrete 75
    ground floors 73–4
    upper floors 75
footprints, buildings 102
formulae, estimating quantities of resources 51–6
formwork, concrete 75
foundations 68–73
    additives 72–3
    aggregates 72
    cavity walling 182–3
    cement 72
    concrete 72
    condensation 182–3
    materials 72–3
    pad foundations 70
    pile foundations 70, 72
    purpose 68
    raft foundations 70, 71
    reinforcement 73
    selecting 71–2
    shear failure 68
    solid walling, foundation concrete 142
    stone, artificial/natural 73
    strip foundations 69, 71
    subsoils 71
    types 69–70
    water 72
framed buildings, masonry cladding 213–22
frog, brick term 139

G

gable end roofs 79
gable ends, cavity walling 192–4
geothermal ground heat 92
goggles 32
ground floors 73–4
ground work 96

H

hand protection 32
handling materials 22–6
HASAWA *see* Health and Safety at Work etc. Act
hazards 5
    creating 17–18
    identifying 13–18
    method statements 14–15
    reporting 16–17
    risk assessments 14–15
    setting out 96–7
    types 15–16
head protection 32
header face, brick term 139
heads, frame, openings 202
health and safety, thin joint masonry 226–7
Health and Safety at Work etc. Act (HASAWA) 2, 5
Health and Safety Executive (HSE) 6, 7
    work method statements 226

health risks 21
hearing protection 20, 21, 32
heat sink systems 91
height, working at 4, 16, 17, 200–1
    equipment 26–7, 200–1
hipped end roofs 79
horizontal 106
housekeeping 14
HSE *see* Health and Safety Executive
HVAC (Heating, Ventilation and Air-conditioning) 83, 84
hygiene 18–21

I

improvement notices 6
inaccuracies/discrepancies, setting out 98
inaccurate estimates 58
industrial standards
    solid walling 156
    thin joint masonry 232
information 40–51
    see also communication
    abbreviations 50
    conformity 49
    drawing equipment 50
    drawings 41–4
    manufacturers' technical information 47
    organisational documentation 48
    plans 41–4
    policies 46
    procedures 46
    programmes of work 44–5
    scales 51
    schedules 47
    specifications 46, 49–50
    training and development records 49
    types 40
information sources, setting out 98
infrastructure 83
injuries 7, 9
insulation 93
    cavity walling 179, 182, 196–7
internal walls 77–8
interpreting specifications 49–50
interviews 82
invert levels 99
isometric projection 44

J

jambs, openings 203
jointing and pointing, thin joint masonry 233–4
joints, movement, solid walling 159

L

labour rates 57
ladders 26–7
landfill 84
lean-to roofs 79
learning skills, solid walling 143
legislation
    health and safety 2–8
    personal protective equipment (PPE) 33
leptospirosis 20, 21
letters 65
lifting, safe 22–3
lighting 93
lime, solid walling 133
lintels
    cavity walling 180, 202–3
    openings 202–3
location drawings 41–2

M

major injuries 7, 10
Manual Handling Operations Regulations 4
manufacturers' technical information 47
mark-up 58
masonry cladding
    brick slips 216–17
    Brikloc cladding system 217
    concrete buildings 216–22
    fire barriers 222
    fixings 215–17
    framed buildings 213–22
    steel-framed buildings 216–22
    timber-framed buildings 216–22
    tools/equipment 215–17
materials, purchasing systems 57
measurements
    dimensional accuracy 98, 113, 121
    estimating quantities of resources 51–6
meetings 65
method statements, risk assessments 14–15
mono-pitch roofs 79
mortars
    additives 135–6
    admixtures 135–6
    solid walling 132, 135–6
movement joints, solid walling 159

N

near misses 5, 10, 11
noise 20

O

openings
    cavity walling 201–3
    heads, frame 202
    jambs 203
    lintels 202–3
    pads/slips 202
    reveals 173, 203
    reverse bonds 173
    setting out 201
    sills, frame 202
    solid walling 173–5
optical site square, right-angle corners 112
organic substances 88
organisational documentation 48
orthographic projection (first angle) 44
over 7-day injuries 7
oversite concrete 99

P

pad foundations 70
pads/slips, openings 202
partitions, walls 77–8
party wall junctions, cavity walling 184–6
PAT *see* Portable Appliance Testing
performance reviews 65
perpendicular 78, 79
personal hygiene 20–1
Personal Protection at Work Regulations 4
personal protective equipment (PPE) 4, 31–3
    legislation 33
    thin joint masonry 227–8
piers
    cavity walling 204–5
    English bond walls, attached piers 164–5
    English bond walls, two-brick hollow isolated pier 167–9
    English bond walls, two-brick solid isolated (detached) pier 166–7

Flemish garden wall bond, attached piers 165–6
Flemish garden wall bond, attached piers and movement joint 169–70
solid walling, half-brick wall with attached pier and raking cut 171–2
pile foundations 70, 72
plans 41–4
pointing, thin joint masonry 233–4
policies 46
Portable Appliance Testing (PAT), electricity 28–9
positioning, solid walling 137
PPE *see* personal protective equipment
preparing construction sites *see* setting out
procedures 46
profiles, corner profiles 118–19
programmes of work 44–5
prohibition notices 6
protective clothing 32
Provision and Use of Work Equipment Regulations (PUWER) 3–4
purchasing systems, materials 57
purlin (exterior roof feature) 80
PUWER (Provision and Use of Work Equipment Regulations) 3–4

Q

quantities of resources, estimating 51–8
    solid walling 131–2
quick set level, transferring levels 116–18
quoin header/stretcher, brick term 140
quotes 58, 63

R

raft foundations 70, 71
reclaiming materials 101
rectangular buildings, setting out 122–5
regulations, health and safety 2–8
reinforcement
    foundations 73
    solid walling 158
Reporting of Injuries, Diseases and Dangerous Occurrences Regulations (RIDDOR) 2, 8, 9–10
resources
    checklist 103–6
    estimating quantities 51–8
    finite/renewable 85–8
    setting out 103–7
    solid walling 130–6
respiratory protection 32, 33
return and rusticated corner, cavity walling 189–92
return angle, brick term 140
reveals, openings 173, 203
reverse bonds, openings 173
RIDDOR (Reporting of Injuries, Diseases and Dangerous Occurrences Regulations) 2, 8, 9–10
ridge (exterior roof feature) 80
right-angle shaped masonry structures 107–25
    3:4:5 method 110–11
    building line 107–9
    corner profiles 118–19
    dimensional accuracy 113
    optical site square 112
    site datum point 109
    transferring levels 113–18
risk assessments 14–15
    preparing construction sites 99
risks 5
    health risks 21
roofs 78–82
    alternative building methods 88
    coverings 80–2
    exterior features 80
    types 78–9, 88

S

safety notices 36–7, 65
sand, solid walling 133–4
scaffold 26–7
scales, drawings 51
schedules 47
    solid walling 129
sectional drawings 42–3
sectional joint, brick term 140
semi-circular arch, cavity walling 205–8
service providers 83
services 83
    existing 101
setting out 96–8
    cavity walling 201
    openings 201
    preparing construction sites 99–103
    rectangular buildings 122–5
    resources 103–7
    solid walling 137–43
    thin joint masonry 230–2
shear failure, foundations 68
signs 36–7, 65
sills, frame, openings 202
site clearance 100, 102–3
site datum points 109, 232
site plans 41
solar thermal hot water system 91
solid floors 74
solid walling 77, 127–76
    adhesives 136
    arches 174–5
    bonding 138–42
    cement 132–3
    concrete coping 160–4
    coping 158
    damp-proof barriers 155–6
    decorative corners 157
    discrepancies/inaccuracies 130
    doors 173–5
    dry bonding 144
    English bond walls 143–5, 147–50, 157–9, 160–4, 164–5, 166–7, 167–9
    English garden wall 151–4
    Flemish bond return corner 146–7
    Flemish garden wall bond 154–7, 165–6, 169–70
    foundation concrete 142
    half-brick wall with attached pier and raking cut 171–2
    industrial standards 156
    learning skills 143
    lime 133
    mortars 132, 135–6
    movement joints 159
    openings 173–5
    positioning 137
    preparation 128–37
    quantities of resources, estimating 131–2
    reinforcement 158
    resources 130–6
    sand 133–4
    schedules 129
    setting out 137–43
    specifications 129, 143–62
    squint corners 157–9
    T-junction, English bond walls 147–50
    terminology 138–42
    tools/equipment 136–7
    toothing 146
    vertical movement joints 159
    water 135

weatherproofing 157
windows 173–5
specifications 46
cavity walling 181–4
interpreting 49–50
solid walling 129, 143–62
square class quoin, brick term 140
squint coin, brick term 140
squint corners, solid walling 157–9
steel-framed buildings, masonry cladding 216–22
stone, artificial/natural, foundations 73
stopped end, brick term 140
storing materials 22–6, 31
straight edge and spirit level, transferring levels 114–15
stretcher face, brick term 139
strip foundations 69, 71
sub-contractors 6
subsoils, foundations 71
sustainability 84–93
alternative building methods 87–8
alternative energy sources 91–3
architecture 85–6
building design 85–6, 89, 93
Building Regulations 89
energy efficiency 89–93
insulation 93
resources, finite/renewable 85–8

T
T-junction, brick term 140
T-junction, solid walling
English bond walls 147–50
Flemish bond 150–1
tanking, damp-proof barriers 156
technical information, manufacturers' 47
tenders 58, 63
terminology
brick 138–42
solid walling 138–42
thin joint masonry 223–44
block wall with return and opening 236–9
cavity wall with a return 242–3
cavity walling 240–3
components 235
Concrete Block Association 226
constructing buildings 232–5
damp-proof barriers 232
drawings 225–6
health and safety 226–7
industrial standards 232
jointing and pointing 233–4
materials 228
personal protective equipment (PPE) 227–8
pointing 233–4
setting out 230–2
weather effects 229

tile, exterior roof feature 80
timber floors 74
timber-framed buildings
masonry cladding 216–22
rusticated corner 219–21
timber framed walls 77
toolbox talks 7, 8, 19
tools/equipment
masonry cladding 215–17
solid walling 136–7
toothing, solid walling 146
training and development records 49
transferring levels
quick set level 116–18
right-angle shaped masonry structures 113–18
straight edge and spirit level 114–15
transverse joint, brick term 140

U
upper floors 75
utilities 83

V
VAT (Value Added Tax) 63, 64
vertical 106
vertical movement joints
cavity walling 198–200
solid walling 159
voussoirs, arches 174

W
walkover surveys, preparing construction sites 100
wall joint, brick term 140
wall ties, cavity walling 180, 198–200
walls 75–8
see also cavity walling; solid walling
internal 77–8
partitions 77–8
timber framed 77
waste control 25–6
water
foundations 72
solid walling 135
weather effects
bricklaying 217
thin joint masonry 229
weatherproofing, solid walling 157
welfare facilities 18–19
wind turbines 92
windows, solid walling 173–5
Work at Height Regulations 4
work method statements, Health and Safety Executive (HSE) 226
work programmes 44–5
work sizes, bricks 139
working platforms 26–7
working space, cavity walling 181

ACKNOWLEDGEMENTS

The author and the publisher would also like to thank the following for permission to reproduce material:

Images and diagrams

AIM Ltd: 7.16; **Alamy**: Arcaid Images: 3.15, blickwinkel: 3.16, Chris Pancewicz: 4.3, fotofacade.com: 6.32, Jonathan Howell Archive: 7.2, Peter Davey: chapter 1 opener; **BSA**: 2.4 and 4.5; **Energy Saving Trust © 2013**: 3.31; **Fotolia**: 1.1, 1.2, 1.3, 1.5, 1.6, 1.7, 1.8, 1.14, 1.15, 1.16, 2.26, 4.2; **Helfen**: 2.3; **instant art**: table 1.15; **iStockphoto**: 1.11, 2.27, 3.7, 3.19, 3.21, 3.22, 3.24, 3.25, 3.28, 3.29, 3.3, 5.57, 5.98, 7.3, 7.4; **J.E. Johnston**: 5.1; **Nelson Thornes**: 1.9, 1.10, 1.12, 1.13, chapter 4 opener, 4.9, 4.10, 4.12, 4.29, 4.30, 4.31, 4.32, 4.33, 4.34, 4.35, 4.36, 4.37, chapter 5 opener, 5.2, 5.3, 5.4, 5.23, 5.25, 5.27, 5.28, 5.29, 5.30, 5.31, 5.32, 5.33, 5.34, 5.37, 5.39, 5.40, 5.41, 5.42, 5.43, 5.44, 5.45, 5.46, 5.48, 5.54, 5.55 5.56, 5.58, 5.61, 5.62, 5.63, 5.64, 5.65, 5.66, 5.67, 5.79, 5.80, 5.82, 5.84, 5.85, 5.86, 5.87, 5.88, 5.89, 5.91, 5.93, 5.94, 5.95, 5.96, chapter 6 opener, 6.4, 6.5, 6.6, 6.8, 6.9, 6.10, 6.12, 6.13, 6.14, 6.15, 6.16, 6.18, 6.19, 6.20, 6.21, 6.22, 6.23, 6.24, 6.25, 6.41, 6.45, 6.46, 6.47, 6.48, 6.49, 6.5, 6.51, 6.52, 6.54, 6.55, 6.56, 6.57, 6.58, 6.59 6.60, 6.61, 6.62, chapter 7 opener, 7.5, 7.6, 7.7, 7.8, 7.9, 7.10, 7.11, 7.12, 7.13, 7.14, 7.15, chapter 8 opener, 8.1, 8.3, 8.12, 8.13, 8.14, 8.15, 8.16, 8.17, 8.18, 8.19, 8.20, 8.21, 8.22, 8.23, 8.24, 8.25, 8.26, 8.27, 8.28, 8.31; **Contains Ordnance Survey data © Crown copyright and database right 2013**: 4.4; **Peter Brett**: 2.2, 2.5, 2.6; **Science Photo Library**: Peter Gardiner: 1.4; **Shutterstock**: chapter 2 and 3 opener, 3.14, 3.17, 3.23, 3.26, 3.27, 6.42, 7.1; **STA**: 5.16; **Wikipedia**: 3.18.

Every effort has been made to trace the copyright holders but if any have been inadvertently overlooked the publisher will be pleased to make the necessary arrangements at the first opportunity.